OCEANS

OCEANS

The Illustrated Library of the Earth

CONSULTING EDITORS

ROBERT E. STEVENSON, Ph.D.

FRANK H. TALBOT, Ph.D.

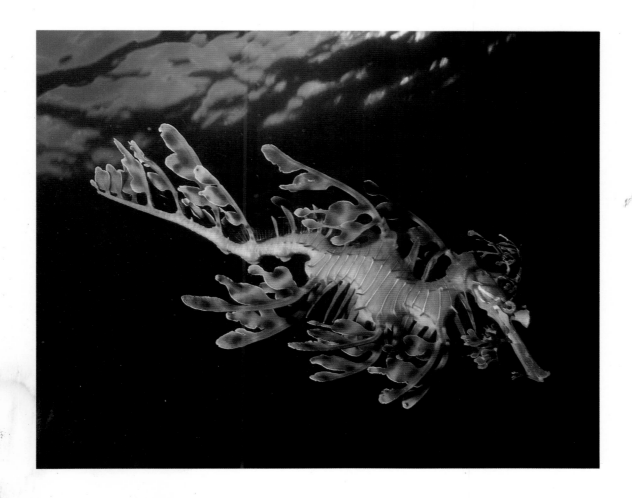

RODALE PRESS
EMMAUS, PENNSYLVANIA

Published 1993 by Rodale Press, Inc.
33 East Minor Street, Emmaus, PA 18098, USA

By arrangement with Weldon Owen
Conceived and produced by Weldon Owen Pty Limited
43 Victoria Street, McMahons Point, NSW, 2060, Australia
Fax (61 2) 929 8352
A member of the Weldon International Group of Companies
Sydney ● San Francisco ● London
Copyright © 1993 Weldon Owen Pty Limited

CHAIRMAN: Kevin Weldon
PRESIDENT: John Owen
GENERAL MANAGER: Stuart Laurence
PUBLISHER: Sheena Coupe
SERIES COORDINATOR: Tracy Tucker
ASSISTANT EDITOR: Veronica Hilton
COPY EDITOR: Sheena Coupe
DESIGNER: Toni Hope-Caten
DESIGN CONCEPT: Andi Cole, Andi Cole Design
COMPUTER LAYOUT: Paul Geros, Veronica Hilton
PICTURE RESEARCH: Annette Crueger, Brigitte Zinsinger
ILLUSTRATION RESEARCH: Joanna Collard
CAPTIONS: Margaret Atkinson, Terence Lindsey
INDEX: Diane Harriman
ILLUSTRATIONS AND MAPS: Jon Gittoes, Mike Gorman, Oliver Rennert
COEDITIONS DIRECTOR: Derek Barton
PRODUCTION DIRECTOR: Mick Bagnato
PRODUCTION COORDINATOR: Simone Perryman

Library of Congress Cataloging–in–Publication Data

Oceans/consulting editors, Robert E. Stevenson, Frank H. Talbot.
   p. cm.—(The Illustrated library of the earth)
   Includes index.
   ISBN 0–87596–596–2 hardcover
   1. Oceanography. I. Stevenson, Robert E. II. Talbot, Frank.
   III. Series.
   GC11.2.026 1993
   551.46—dc20   92–46129
                                        CIP

If you have any questions or comments concerning this book, please write to:
Rodale Press
Book Readers' Service
33 East Minor Street
Emmaus, PA 18098

Production by Mandarin Offset, Hong Kong
Printed in Hong Kong

Distributed in the book trade by St. Martin's Press

10 9 8 7 6 5 4 3 2 1

A WELDON OWEN PRODUCTION

JACKET: *An aerial view of a section of the Great Barrier Reef, Australia. Photo by Michael Jensen/AUSCAPE.* ENDPAPERS: *Small marine fish often pack tightly together for mutual protection: these are glassy sweepers* Parapriacanthus guentheri *in the Red Sea. Photo by David Doubilet.* JACKET INSET & PAGE 1: *Gaudy butterfly fish among coral in the Red Sea. Photo by Lionel Isy-Schwart/The Image Bank.* PAGE 2: *Soldier fish or squirrel fish feed at night and rest quietly in schools during the day. Photo by Pam & Willy Kemp/Oxford Scientific Films.* PAGE 3: *The sea dragon* Phycodurus eques *is common wherever seaweed occurs. Photo by David Doubilet.* PAGES 4–5: *A moray eel emerges from its lair amongst orange coral.* PAGES 6–7: *A pod of humpback whales feeding, Alaska.* PAGE 8: *A clownfish sheltering in sea-anemones. Photo by Australian Picture Library/ZEFA.* PAGES 10–11: *King penguins* Aptenodytes patagonicus *march from the surf at Macquarie Island.* PAGES 12–13: *Two divers explore the world beneath the sea's restless surface.* PAGES 52–53: *A king angelfish, Sea of Cortez, Mexico.* PAGES 104–105: *Looking lean and lethal in silhouette, a school of barracudas swirls around a diver in New Guinea waters.*

David Doubilet

CONTRIBUTORS

MARGARET ATKINSON
School of Biological Sciences, University of Sydney, Australia

DR. MICHEL A. BOUDRIAS
Department of Zoology, University of San Diego, USA

PROFESSOR ALEC C. BROWN
Department of Zoology, University of Cape Town, South Africa

PROFESSOR MICHAEL BRYDEN
Department of Veterinary Anatomy, University of Sydney, Australia

DR. M.G. CHAPMAN
Institute of Marine Ecology, University of Sydney, Australia

DR. SYLVIA A. EARLE
Deep Ocean Engineering Inc., USA

DR. SCOTT C. FRANCE
Woods Hole Oceanographic Institution, Massachusetts, USA

DR. CHRISTIAN D. GARLAND
Departments of Geography and Environmental Studies, and Agricultural
Science, University of Tasmania, Australia

DR. STEPHEN GARNETT
Queensland Department of Environment and Heritage, Australia

DR. RICHARD HARBISON
Woods Hole Oceanographic Institution, Massachusetts, USA

PROFESSOR ROBERT R. HESSLER
Scripps Institution of Oceanography, University of California,
San Diego, USA

DR. ANGELA MARIA IVANOVICI
Australian National Parks and Wildlife Service, Australia

DR. JAMES C. KELLEY
Department of Science and Engineering,
San Francisco State University, USA

DR. KNOWLES KERRY
Australian Antarctic Division, Australia

DR. G.L. KESTEVEN
formerly Food and Agriculture Organization

DR. M.J. KINGSFORD
School of Biological Sciences, University of Sydney, Australia

DR. JOHN E. McCOSKER
Steinhart Aquarium, California Academy of Sciences, San Francisco, USA

DR. COLIN MARTIN
Department of Scottish History, University of St. Andrews, UK

DR. JOHN R. PAXTON
Australian Museum, Sydney, Australia

PROFESSOR VICTOR PRESCOTT
Department of Geography, University of Melbourne, Australia

DR. PAUL SCULLY-POWER
Scripps Institution of Oceanography, University of California,
San Diego, USA

DR. J.R. SIMONS
formerly Faculty of Science, University of Sydney, Australia

DR. ROBERT E. STEVENSON
International Association for the Physical Sciences of the Ocean,
Del Mar, USA

PROFESSOR A.J. UNDERWOOD
Institute of Marine Ecology, University of Sydney, Australia

DR. DIANA WALKER
Department of Botany, University of Western Australia, Australia

CONTENTS

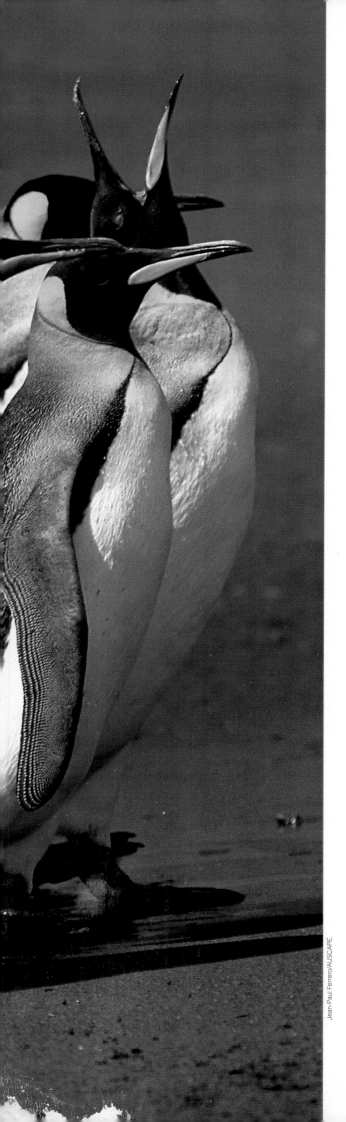

Jean-Paul Ferrero/AUSCAPE.

FOREWORD

Though the oceans make our planet unique, we still know very little about how they work or what is in them, and we have explored only a fraction of the deep-sea floor. The ocean has always been a mystery and a challenge, the source of a mythology of monsters, mermaids, and drowned cities. Knowledge has come slowly: oceanography, the science of the oceans, is little more than a century old, beginning with a systematic study by HMS Challenger in the 1870s. Its naturalists described a myriad of strange creatures brought up from the depths by dredges and nets.

Over a hundred years earlier divers had begun clumping across the shallow sea bottom wearing lead boots and hoses with compressed air, looking for sponges and pearls, but only since the 1930s have humans been able to go below the sunlit upper layers to the "twilight zone" and deeper. The first was William Beebe, who was lowered into the waters near Bermuda in a steel bathysphere and through its thick window made excited notes and sketches of strange midwater creatures. In the 1940s, Jacques Cousteau and Emile Gagnan designed the valve which led to scuba and opened up the shallow-water reef zones to personal exploration. In 1960 Jacques Piccard and Don Walsh reached the deepest spot on Earth at the base of the Mariana Trench.

Today we are able to scan the sea from space to help us understand temperature, water level, and currents. We can pull up cores of bottom sediments to provide clues about the history of the sea floor. Only a few decades ago the submersible Alvin found a whole new ecological system on the deep-sea floor —a system based on chemical energy released by bacteria from heated, sulfide-rich water pouring out of bottom fissures.

Huge areas of the deep oceans are still to be traveled and it is certain that we will find unusual physical structures, thousands of new species, new usable resources, and perhaps even different geological and biological processes. There is another reason to learn more, and that is that our present usage is not always wise, and many estuaries, coastal waters, and enclosed seas are now polluted. We are also overfishing, etching into the capital of the oceans, and not just using their surplus bounty. We know, too, that the oceans are important to changes in the composition of the atmosphere. There is an urgency to explore and to understand the oceans . . . for they form part of our future.

FRANK TALBOT
Director, National Museum of Natural History,
Smithsonian Institution, Washington DC, USA

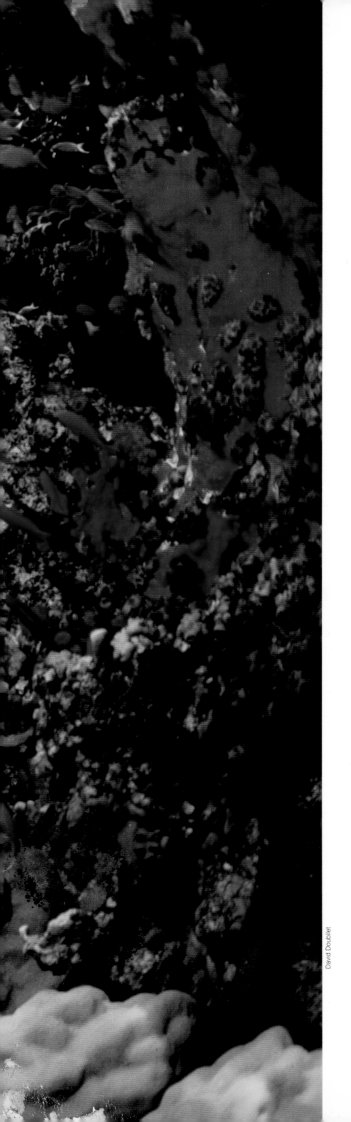

David Doublet

1 THE UNDERSEA WORLD

ROBERT E. STEVENSON

O ur *"Blue Planet", so named because of its distinctive appearance from outer space, derives its brilliant coloration from the vast oceans that cover 70 percent of its surface. One can begin to appreciate the immense size of the oceans by considering that they contain over 450 billion liters (100 billion gallons) for every inhabitant of the Earth. The world ocean, a vital natural resource, is the result of ancient geological processes that have created extraordinary undersea landscapes of ridges and trenches, mountain chains, valleys, and plains.*

FORMING THE OCEANS

The oceans were formed over the millennia from steam given off from the interior of the Earth by the action of volcanoes, and they acquired their salty composition by the continual weathering and leaching of the rocks through countless cycles of evaporation and precipitation. Sea water is as a result 3.5 percent by weight dissolved salts, a percentage that remains constant within very narrow margins throughout the oceans.

Although settled communities of farmers had appeared along the fertile valleys of Mesopotamia many thousands of years earlier, there was no concept of the global expanse even until 400 BC when Aristotle was puzzled by marine fossils in rocks high above sea level. In the centuries that followed, the history of the Earth continued to puzzle scholars to the extent that in 1654 Archbishop James Ussher decided to settle the question once and for all by declaring that Heaven and Earth had been created "upon the entrance of the night preceding" Sunday October 23, 4004 BC.

THE BEGINNINGS OF UNDERSTANDING

Ussher's dogmatic pronouncement was not acceptable to the scientists of the day, but the power of the Church was such that it would take another 200 years to overcome this Church-dominated concept. In the mid-nineteenth century James D. Dana, an American geologist, declared that the Earth had originated in a molten state, that it had contracted upon cooling, and that it continued to do so. The continents and ocean basins, he declared, were permanent features, having stabilized at the beginning of geologic time. The continents had cooled first, and further contractions had lowered the crust to form the oceans. As the cooling continued, the Earth's interior had shrunk, causing tremendous forces that created the mountain ranges—much like the wrinkles on a dried apple.

Dana's concept was widely adopted and geological research into the twentieth century was largely based on his theory. Then, after half a century of speculation, and an extraordinary decade of oceanographic expeditions in the mid-century, the origin of the Earth's present continents and ocean basins was finally known. The long and controversial efforts to prove or disprove that moving, crustal plates formed the ocean basins and continents finalized in 1967; the process is known as plate tectonics.

CONTINENTAL DRIFT

At the start of the twentieth century there were no topographical maps of the sea floor. Geologists believed the Earth was slowly cooling and shrinking, and that continents and oceans had been in their respective places since the beginning of time. With this prevailing view, we can understand why Dr. Alfred Wegener's theories describing continents that drift or spread apart were not taken well by his fellow German colleagues when in January 1912 Wegener presented his ideas on "The Formation of Major Features of the Earth's Crust (Continents and Oceans)". In 1915, his theory of continental drift was published in a small but comprehensive book entitled *The Origin of Continents and Oceans*.

Wegener described how continents had once been united, but had drifted to their present locations in the past millions of years. It was a well-conceived hypothesis. He pointed out that continental boundaries lay at the edge of the surrounding shelves rather than the present coastlines. The oceanic crust was similar

Giraudon, Paris

Opposite. Elemental forces clash as molten lava from a volcanic eruption in Hawaii flows into the pounding waves, producing huge clouds of steam. Though volcanic in origin, the Hawaiian islands are not part of the "ring of fire" that rims the Pacific.

One of the earliest world maps. Drawn by the Greek geographer Ptolemy around AD 140, it portrays the Earth's surface as mostly land. The realization that the reverse is true came only gradually: the process started with the travels to the orient of Nicolo de' Conti, reported in 1444, and Portuguese seafaring explorations directed by Prince Henry the Navigator from 1424, and culminated in the first global circumnavigation by Ferdinand Magellan in 1522.

STAGES IN CONTINENTAL DRIFT

200 million years ago

70 million years ago

130 million years ago

Today

Over some 200 million years the primordial continent Pangaea gradually broke up into Laurasia and Gondwana, which in turn fragmented to form the modern continents about 50 million years ago.

The continents ride some half-dozen major tectonic plates, whose boundaries are marked by volcanoes and zones of maximum earthquake activity.

to pitch, he suggested, which flows when put under lengthy pressure, yet breaks into brittle pieces when struck a blow. Fossil and geological evidence indicated former connections between continents. By examining evidence of this kind, Wegener concluded that the major landmasses had been joined in the geologic past and had since drifted apart.

Wegener named his ancient supercontinent Pangaea, from the Greek root meaning "all land". He assumed that Pangaea had existed 300 million years ago, and began to separate about 200 million years before the present, with the continents slowly and imperceptibly

moving to the positions they occupy today. Geologists around the world were stung by the impertinence of Wegener's proposition. From a special meeting of the Royal Geographical Society in London in January 1923, Wegener learned that no one who "valued his reputation for scientific sanity" would advocate a wild theory like continental drift. Nevertheless, Wegener continued to pursue the theory until his death in 1930. The data and technology needed to prove, or disprove, the concept of continental mobility would remain the secret of the ocean basins for another 30 years.

EXPANDING HORIZONS

In 1950, ocean-testing equipment was more sophisticated than in Wegener's day. Acoustic depth recorders enabled scientists to determine depths for mapping the ocean floor; sensitive heat probes and radioactive carbon-dating techniques were used to analyze the age and polarity of the layers of the sea floor; and core sea-floor samples told its biological history. After the Second World War, surplus naval vessels equipped with these new instruments were made available to United States oceanographers. Scientists from Scripps Institution of Oceanography, California, sailed into the Pacific Ocean, and other exploring cruises were mounted from Woods Hole Oceanographic Institution, Massachusetts, and the Lamont-Doherty Geological Laboratory, New York.

The data collected on these expeditions challenged the assumptions that were based on the prevailing geologic thought. By 1960, the three major ocean

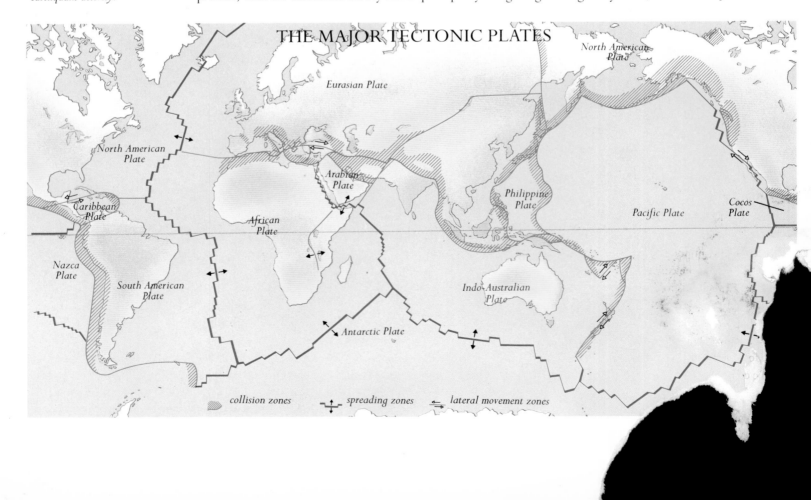

THE MAJOR TECTONIC PLATES

North American Plate

Eurasian Plate

North American Plate

Arabian Plate

Philippine Plate

Pacific Plate

Cocos Plate

Caribbean Plate

African Plate

Nazca Plate

South American Plate

Indo-Australian Plate

Antarctic Plate

collision zones · spreading zones · lateral movement zones

basins and their adjacent seas had produced some confirmations, some disproofs, and an extraordinary amount of new data leading to new avenues of thought and controversy among scientists.

The implication that the sea floor was moving, and had been doing so for the past 200 million years, presented a puzzling enigma to most geologists of the day. Not so to Dr. Robert S. Dietz in California, nor to Professor Harry Hess at Princeton University. Simultaneously, they considered the new knowledge and proposed that the ceanic crust is continuously spreading from the oceanic . Dietz called the concept "sea-floor spreading". r many expeditions to the Atlantic and Pacific oceans, ears of data analysis, Dietz and Hess established the e of con ection currents—the powerful force the movement of continents. f)or spreading in 1960 was as incredible to the ommunity as continental drift had been in 1912. ing data, even non-believers came ss claim. From Cambridge

University came Dr. Drummond H. Matthews and his student, Frederick J. Vine, who tested the hypothesis by measuring the magnetic anomalies in the oceanic crust on either side of the ridges in the Atlantic and Indian oceans. By 1963, they had confirming data: the sea floor is spreading from the ridge crests and at rates that agree with the 200-million-year period starting with the break-up of Pangaea.

With sea-floor spreading accepted, the next question arose: Do the oceans' bottoms spread as two massive units in each basin? Not likely, thought Dr. J. Tuzo Wilson, a Canadian geologist on sabbatical at Cambridge University, in 1965. He wrote, "These huge transverse faults are not isolated features, but connect with the ridges and trenches to divide the oceanic crust into large, rigid plates." This made sense to Dr. Jason Morgan, a young geologist, who in 1967 laid the last piece of the puzzle before his colleagues in Washington, DC by demonstrating how all the Earth's plates are spreading and rotating around their individual poles.

An image from the space shuttle Columbia *views the distant Mediterranean Sea from high above the Red Sea. The Sinai Peninsula at center is defined by the Gulf of Suez on the left and the Gulf of Aqaba on the right; part of the great African Rift can be seen extending upward through the Dead Sea toward Turkey. Five million years ago the Red Sea was a shallow basin. Now widening, as the sea floor spreads to separate the Saudi peninsula from continental Africa, it is considered to be an embryonic ocean.*

THE EARTH'S DYNAMIC CRUST

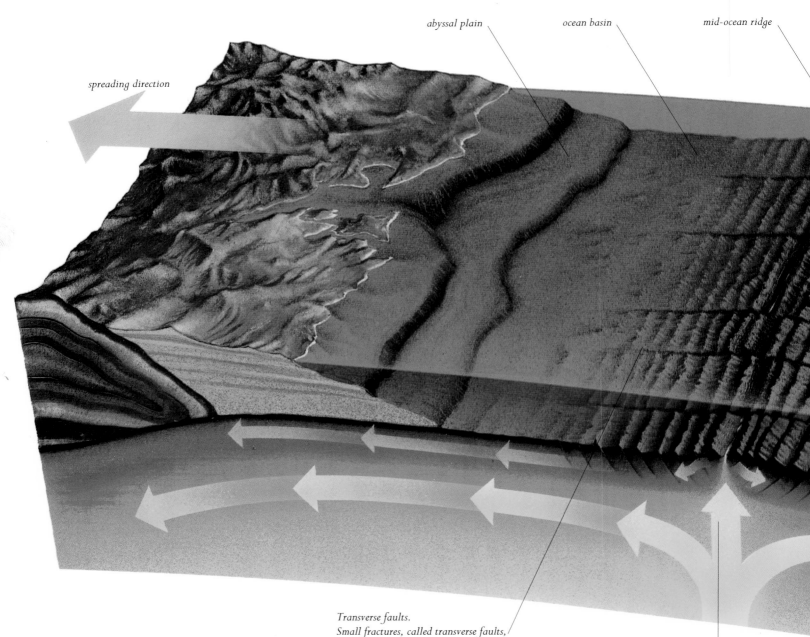

abyssal plain

ocean basin

mid-ocean ridge

spreading direction

Transverse faults.
Small fractures, called transverse faults,
form across spreading centers between newly
formed plate segments that separate at
different rates.

Magma upwelling.
Molten magma seeps upward at cracks in the fracture
zones along the deep ocean floor, forming ridges as the
material solidifies. The sea floor is forced apart, slowly
spreading to form oceanic plates on either side.

Convection forces deep within the
Earth's mantle govern the slow
movements of the outer crust on
which the continents ride, shuffling
them around at a rate of about
6 centimeters (2 inches) per year
—the growth rate of a fingernail.
Three major structural (or
"tectonic") features are involved in
these movements: oceanic ridges,
mountains and trenches, and
transverse faults.

STAGES IN SEA-FLOOR SPREADING

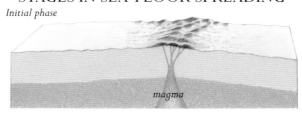

Initial phase

magma

Developing phase

rift valley

spreading direction

spreading direction

magma

THE OCEAN BASINS

Although they are interconnected, the three major ocean basins of the world—the Pacific, the Atlantic, and the Indian—have distinctive outlines. The smaller marginal seas of the Gulf of Mexico, the Caribbean Sea, the Mediterranean Sea, and the North Polar Sea, as well as the marginal basins landward of the island arcs off eastern Asia and Australia, all have pronounced barriers that partially separate them from the oceans.

THE PACIFIC BASIN

The Pacific, the largest and deepest of the oceans, covers 166 million square kilometers (64 million square miles) and has a mean depth of 4,188 meters (13,737 feet). The circular Pacific lacks the symmetry of the Atlantic, which curves proportionally with the landmasses that surround it. In contrast, the eastern Pacific, north of the tip of Baja California, is dominated by long east–west fracture zones that intersect North America at right angles. South of Baja,

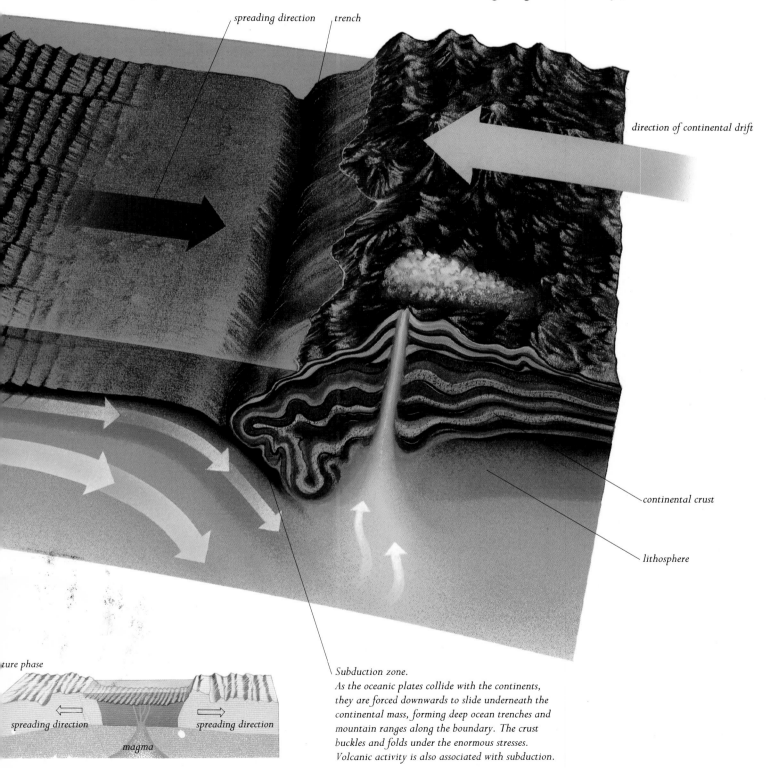

spreading direction

trench

direction of continental drift

continental crust

lithosphere

ture phase

spreading direction

spreading direction

magma

Subduction zone.
As the oceanic plates collide with the continents, they are forced downwards to slide underneath the continental mass, forming deep ocean trenches and mountain ranges along the boundary. The crust buckles and folds under the enormous stresses. Volcanic activity is also associated with subduction.

The mineralized stack of this black smoker (hydrothermal vent) on the East Pacific Rise spews out an inky stream of steaming water and hydrogen sulfide. Material from the fountain is utilized by the sulfur-oxidizing bacteria that are found in many of the animals inhabiting the areas around hydrothermal vents.

Opposite. *Lava erupts in a fiery fountain and flows as incandescent* pahoehoe *from the main vent of Hawaii's Kilauea, the largest active volcano on Earth. Around the world, the pattern of active volcanoes and most intense earthquake activity conforms closely with the boundaries of the major tectonic plates.*

the East Pacific Rise and its fractures bend gently westward, cross the south Pacific and merge with the Indian Ocean Ridge south of Australia.

The sea floor of the central Pacific, both north and south, is made up of a number of ridges that trend northwest–southeast. These seamount chains are crowned in many places with islands: Hawaii in the north Pacific, and the coral atoll belts of the tropical and southwest Pacific. The island arcs and their associated trenches ring the Pacific basin from southernmost New Zealand to the tip of Tierra del Fuego. The largest gap in this ring of trenches is between British Columbia and the southern end of the Gulf of California.

The East Pacific Rise is part of the mid-oceanic ridge system of the major ocean basins. Its crest rises 2–3 kilometers (1–2 miles) above the sea floor, but in contrast with the Mid-Atlantic Ridge, the East Pacific Rise is thousands of kilometers wide with gradual slopes. The ridges and troughs parallel to the crest do not have the relief of those in the Atlantic, the flanks are covered with volcanoes, and the crest is almost devoid of sediments.

The Gulf of California began forming about 6 million years ago as Baja California split away from the mainland of Mexico. The splitting continues today as Baja moves west at about 6 centimeters (2.5 inches) per year. The East Pacific Rise, the splitting Gulf of California, and the San Andreas Fault in California are interconnected. The East Pacific Rise interacts with the North American continent by the same mechanism that has created the rift valleys of Africa and Israel's Dead Sea and Sea of Galilee!

From the international date line to the Americas, and from Punta Gorda south nearly to the equator, the crust of the Pacific sea floor is ripped by the greatest fractures on Earth. The Clipperton Fracture Zone, for example, extends past the Line Islands, south of Hawaii, and through the Phoenix Islands across the equator. The fracture zones are basically parallel to each other, evenly spaced, with ridges and depressions along their length. Where the fractures cross the East Pacific Rise, the sea floor is displaced in an east–west direction and it is here that great earthquake activity occurs. The Mendocino Fracture Zone has the largest escarpment, the sea floor on the south side being some 1,200 meters (4,000 feet) deeper than on the north.

Although at first glance the seamounts, coral reefs, and volcanic islands of the central Pacific seem to be scattered like so many dobs of mud, they are actually aligned on north–south and northwest–southeast volcanic ridges. The longest is the Hawaiian Ridge, which extends northwest for 4,500 kilometers (2,800 miles) from the island of Hawaii, connecting there with

the Emperor Seamount Chain which continues north until it disappears at the junction of the Kuril and Aleutian trenches. South of Hawaii are the Line Islands, including Kiritimati, the oldest known atoll. South and west of Kiritimati are the great series of coral reefs that make up the Marshall and Caroline islands and Kiribati. South of the equator, atop linear ridges, lie the Tuamotus, Samoa, the Society Islands, and the Austral Ridge. The high islands of all the ridges are volcanic, and borings in the atolls confirm that volcanic material underlies the coral reefs.

The islands and submerged reefs of the Hawaiian Ridge extending to Midway are all volcanic, and studies of the ages of the eruptive rocks indicate that the volcanism has progressed southeast at a rate of about 15 centimeters (6 inches) per year. The implication is that there is a "hot spot" in the oceanic crustal layer and that the sea floor has moved to the northwest over this spot to create the volcanic peaks. The "Big Island" of Hawaii currently sits on a hot spot, evidenced by Kilauea's recent eruptions. In the past decade, scientists have encountered sea-floor eruptions south of the "Big Island", which may indicate that Hawaii is moving off the center of this eruptive hot spot.

To the south and west of Hawaii is a vast area of coral reefs and volcanic peaks. Many of the peaks have flat tops, suggesting erosion by surface-wave action during an earlier geologic time when they were at the sea's surface. These flat-topped peaks, called guyots after the Swiss geologist Arnold Guyot, have average depths of 1,200 meters (3,940 feet). If the guyots were eroded by waves, then the whole central Pacific must have subsided by this amount. But we have no evidence of any such subsidence, or how it may have happened.

The most geologically active structures in the world are the island arcs and their accompanying trenches in the western Pacific, the great trench off Indonesia, and that bordering Peru and Chile. They make up the deepest parts of the ocean basins, and experience the greatest earthquake activity. There are gaps in the trenches adjacent to America, especially along the United States and Canada.

The western Pacific trenches are much deeper than those along the borders of the eastern Pacific. Starting near the Kamchatka peninsula on the eastern fringe of Russia, they form a disconnected series south to the Kermadec Islands north of New Zealand. The similarity of the depths in the trenches is striking. Except for that off New Britain, which bottoms at 8,320 meters (27,300 feet), they are all deeper than 9,000 meters (30,000 feet), with the Mariana Trench taking the "depth prize" at 10,860 meters (35,630 feet), a bare distinction over the Tonga Trench at 10,822 meters (35,505 feet) deep.

THE OCEAN BASINS

T he ocean floor is a dramatic landscape, consisting of mountain ranges, deep trenches, volcanoes, and vast plains. There are three major ocean basins. The Pacific is by far the largest at 166 million square kilometers / 64 million square miles, followed by the Atlantic (86 million square kilometers / 33 million square miles) and the Indian (73 million square kilometers / 28 million square miles).

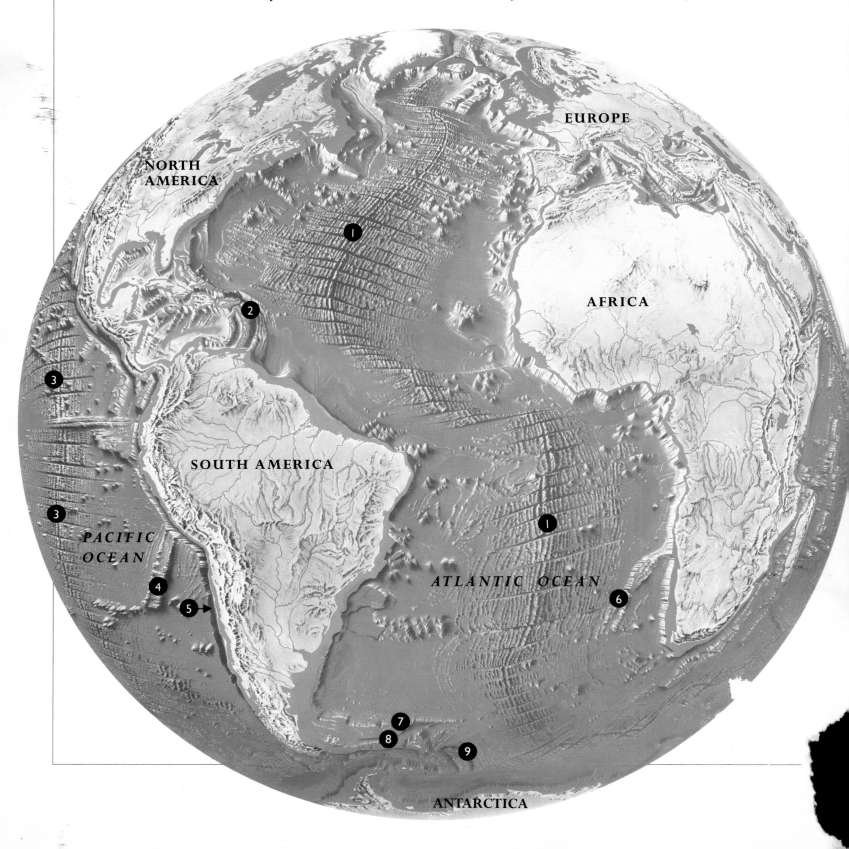

EUROPE

NORTH AMERICA

AFRICA

SOUTH AMERICA

PACIFIC OCEAN

ATLANTIC OCEAN

ANTARCTICA

SIGNIFICANT FEATURES

1. Mid–Atlantic Ridge
2. Puerto Rico Trench
   (9,200 meters/30,184 feet)
3. East Pacific Rise
4. West Chile Rise
5. Peru–Chile Trench
6. Walvis Ridge
7. Falklands Escarpment
8. Scotia Ridge

9. Sandwich Trench
   (8,264 meters/27,106 feet)
10. Aleutian Trench
    (7,822 meters/25,663 feet)
11. Emperor Seamount Chain
12. Kuril Trench
    (10,542 meters/34,587 feet)

13. Japan Trench
    (10,554 meters/34,626 feet)
14. Hawaiian Ridge
15. Arabian Sea Fan (Indus Cone)
16. Bengal Fan (Ganges Cone)
17. Mariana Trench
    (10,860 meters/35,630 feet)

18. Philippine Trench
    (10,497 meters/34,439 feet)
19. Central Indian Ridge
20. Ninetyeast Ridge
21. Java Trench
    (7,450 meters/24,442 feet)
22. Kermadec–Tonga Trench
    (10,822 meters/35,505 feet)

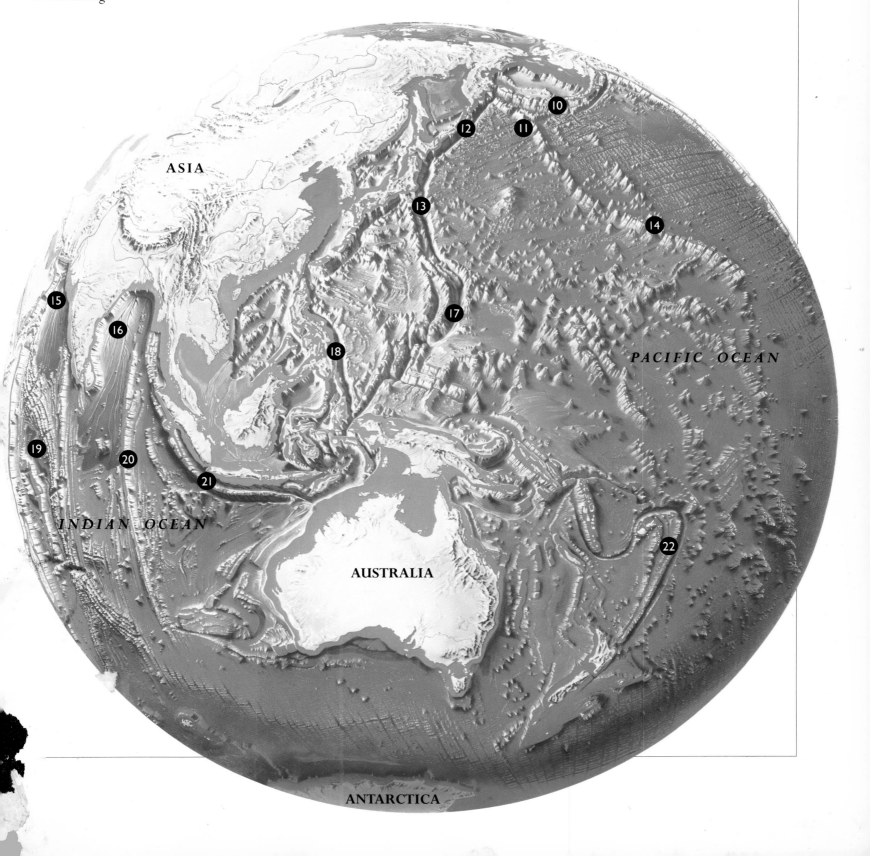

ASIA

PACIFIC OCEAN

INDIAN OCEAN

AUSTRALIA

ANTARCTICA

One of the longest rifts on the Earth's surface is the Great Rift Valley, which extends from the Jordan River south through the Dead Sea, the Ethiopian Danakil depression, Kenya, and Tanzania to the Shire Valley in Mozambique. Extensive volcanism along its length has produced such features as Mount Kilimanjaro (5,895 meters/19,340 feet high) and Lake Rukwa, at 1,436 meters (4,607 feet) the second deepest lake in the world.

Krafft-Explorer

THE ATLANTIC BASIN

The elongated Atlantic, with its curving parallel shores, has an area of 86 million square kilometers (33 million square miles) and a mean depth of 3,736 meters (12,254 feet). The borders of the ocean are remarkably symmetrical, a symmetry mirrored by the Mid-Atlantic Ridge which bisects the basin. The mean relief of the ridge above the basins that flank it is close to 700 meters (2,300 feet), but in some places it stands as much as 3,000 meters (10,000 feet) above an adjacent plain.

Scientists discovered a central ridge in the north Atlantic in 1873. After the First World War, and the development of acoustic depth finders, the ridge was mapped to the south Atlantic. Later, with the use of precision depth recorders, it became clear that the Mid-Atlantic Ridge was part of a world-encircling submarine mountain system. Today we can view this dynamic oceanic ridge system using radar altimetry. The central rift valley of the ridge is the focus of numerous earthquakes, indicative of the tensile forces that pull the oceanic crust apart. At the center of the rift are recent lavas, and in many places hot gases and magma pour out onto the sea floor, constantly forming new rocks, highly mineralized deposits, and nutritive bottom water.

The mid-oceanic ridge is cut by many transverse fractures with deep, elongated valleys that offset the ridge crest. These are epicenters for nearly continuous earthquake activity adjacent to the ridge, and have long histories of volcanism, some of which has produced seamount chains, as off New England, or punched islands to sea level, as in the Azores and St Helena. The fractures sometimes extend into the continents bordering the ocean. The Romanche Fracture, cutting the ridge at the equator, is the deepest at 7,856 meters (25,768 feet). The Falkland Fracture Zone is the longest, extending to the very tip of South Africa and intersecting the continental margin off Argentina through the Falkland Plateau.

Of the two trenches in the Atlantic Ocean, the Puerto Rico Trench is the deeper: 9,200 meters (30,184 feet). Scientists have traced it completely around the volcanic and seismic arc of the West Indies. To the south, the

Pete Turner/The Image Bank

deep-sea boundary of the arc of the Sandwich Trench connects South America with Antarctica. Volcanic and seismic activity is continuous in this trench. It is 8,264 meters (27,106 feet) deep.

THE INDIAN BASIN

The Indian Ocean covers 73 million square kilometers (28 million square miles), and has a mean depth of 3,870 meters (12,700 feet). Its most striking sea-floor feature is an inverted Y-shaped ridge, the counterpart to the Mid-Atlantic Ridge and the East Pacific Rise. One arm of the "Y", the Southwest Indian Ridge, extends around Africa to connect with the Mid-Atlantic Ridge. The other arm, the Mid-Indian Ridge, bends around Australia to join the East Pacific Rise. North of the "Y", the Central Indian Ridge continues north toward the Gulf of Aden and

A dramatic time exposure shows the impressive trajectory of lava and rock fragments in this "explosive" eruption in Iceland. Volcanic activity in Iceland is due to its position at the northern end of the Mid-Atlantic Ridge.

bends into the Gulf, as the Carlsberg Ridge, to join with the rift valleys of Africa and the Red Sea.

Bounding the east side of the Bay of Bengal and its huge sedimentary deposit, the Bengal Fan, is the strikingly straight, north–south, Ninetyeast Ridge. To the west of India is another linear ridge, more or less north–south, atop which lie the Laccadive and Maldive islands. In the western Indian Ocean, northeast of the island of Madagascar, is the unusual Mascarene Plateau. The rocks of this plateau, including the Seychelles Islands, are granitic and of continental origin, giving the plateau the appearance of being a mini-continent amid an ocean basin. South of Madagascar, coral reefs lie on top of volcanic rocks which are clearly part of the ocean

Yet further south, the Agulhas Plateau off South Africa, another mini-continent, is made up of rocks similar to those of the Seychelles.

The Bengal Fan in the eastern part of the Indian Ocean is unique. In no other place has the sea floor been so influenced by deposition from the adjacent land—in this case, from the monumental discharges of the Ganges–Brahmaputra rivers. Extending for 2,000 kilometers (1,250 miles) from the great delta to the south of Sri Lanka, this massive deposit on the sea floor slopes at a nearly even gradient of 1.5 meters per kilometer (8 feet per mile). Numerous branching channels criss-cross the fan, indicating the continuous slow currents that carry the land-derived sediments over its entire length. It is the most gigantic pile of sediments in the world!

KINDS OF CONTINENTAL MARGIN

Wave-beveled and faulted margin

Prograded faulted margin

Fold-beveled inner margin, prograded at outer edge

Prograding margin

Basement rock forms outer sediment dam

Coral reefs form outer sediment dam

Salt domes form outer sediment dam

Volcanic action forms outer sediment dam

Above, right. *Deep Rover at 200 meters (650 feet) in the Bahamas. Development of such submersible vehicles has permitted direct exploration of the oceans, culminating in the descent of the Trieste to 10,860 meters (35,630 feet) on the floor of the Mariana Trench in 1960.*

Opposite. *In many places the continental shelf drops off abruptly to the ocean depths, as in this picture taken off the coast of San Salvador.*

The Continental Margin

Around all the continents and the major islands are shelves that extend from the shore to depths of 100 meters (350 feet) or so. These continental shelves grade into continental slopes that usually meet a continental rise at depths of 3,000–4,000 meters (10,000–13,000 feet). Together, the shelf, slope, and rise make up the continental margin.

HOW SHELVES ARE FORMED

The topography of these submarine features is the result of erosion and deposition during the rise and fall of sea level through the most recent ice ages. When the sea level was at its lowest, some 20 million years ago, the areas that now make up the continental shelves were laid bare of water. Waves cut at a much lower level than today,

David Doubilet

rains and winds carved the shelf surfaces, and rivers flowed across them, building deltas to form the continental slope. As the glaciers melted, the sea rose and gradually covered the shelves, and beaches, barrier islands, and sand flats were formed. Although the beaches became somewhat masked as the sea continued to rise, many are still identifiable, such as those on the shelf off New Jersey on the east coast of the United States.

As depicted on charts, the shelves seem to be flat, with average depths of 60 meters (200 feet) and terminating at depths around 130 meters (430 feet). There are, however, hills, longitudinal and transverse gullies, sand ridges, and sand waves scattered over all the surfaces. The average slope across all shelves is seven minutes of a degree: a slope less than that allowed on the best of billiard tables!

Data from shipborne seismic profilers provide the basis for classifying the structure of the world's continental margins. Though all have been modified by the action related to changing sea levels, wave erosion, changes from currents, and slumping of sediments down the slopes, the underlying structural framework has remained undisturbed.

TYPES OF SHELVES

Earthquake faults in the Earth's crust are common around the continental slopes. Faults are to be expected at the edge of continents, for this is where many earthquakes originate. The shelf inside the faulted slopes may have several origins. Older rocks, folded and faulted, may have been eroded into a rather flat shelf as the sea level rose. The material from the eroded rocks would have slumped onto the faulted slope, or if the

scarps were steep, slid off into deep water. Other shelves have been built along coasts where prodigious supplies of sediments poured out from the adjacent land, forcing the underlying crust to sink continuously from the weight of the deposits. This type of shelf and slope, formed under conditions of subsidence, has a structure similar to those of great river deltas such as the Nile.

The most intriguing shelves are those with a geologic "ridge" at their outer edges. Whether the ridge is part of the basement rock of the continent, a coral reef, salt domes, or a volcano, the projection acts as a dam to catch the sediments from the l... sea level rises. Many shelves have been forme... way. Coral-reef dams occur off the east coast of Fl... and in the Java Sea. On the wide shelf off Texas, sal...

domes make up the dam, and potentially hold major pools of oil. For the most part, volcanic dams—those off the eastern Asiatic coast and in the Arctic—are responsible for the widest shelves. The fill in these huge shelves has been provided by the great rivers entering the sea across the coasts.

SHELF WIDTH AND DEPTH

The average width of continental shelves is 75 kilometers (46 miles). Off the coast of the United States, the width varies from zero at Cape Canaveral, Florida, to just over 200 kilometers (125 miles) off Texas. The 600-kilometer (370-mile) wide shelf of the Bering Sea is the widest off any American coast. Cape Canaveral is probably the only place in the world where there is no shelf: it is unique because of millions of years of erosion by the Gulf Stream. Even where the shelf is narrow, just a few kilometers wide, there is always a slope diving off into thousands of meters of ocean. This occurs off the coasts of California, southern Saudi Arabia, Peru, Chile, and the east coast of India.

Wide shelves have many interesting features. Take, for instance, the shelf off northern Australia. At one end it terminates with New Guinea whereas north of Darwin it suddenly drops off into the Indonesian Trench. Crossing over this huge, deep trench, Sumatra, Java, Borneo, and the Malaysian Peninsula lie atop the shelf and are the most vigorous volcanic shelf islands in the world. The shelf off western Europe contains the entire United Kingdom, plus the North Sea, the Skagerrak, and the English Channel. Other shelves on which islands lie include those that hold Wrangel and the New Siberian Islands and the huge island of Novaja Zemla that separates the Barents Sea from the Kara Sea.

SUBMARINE CANYONS

The existence of canyons cutting deep into the continental margins has been known for more than a century, although their origins and significance in marine processes have been debated. Submarine canyons have steep walls and sinuous valleys with V-shaped cross-sections. Their axes slope outward continuously, and many have reliefs comparable to those of the largest land canyons. Some, such as La Jolla off southern California, have heads that are nearly at the shore. Others, like those off the west coast of Corsica, appear to be recently submerged continuations of land canyons or, perhaps, the inner portion of submarine canyons elevated to become land canyons. At the mouth of the Congo River, a large submarine canyon extends 25 kilometers (15 miles) into an estuary and is 450 meters (1   deep at the bay mouth.

        gical as the association between modern rivers   canyons may seem, there are some canyons that   ear to be related to ancient rivers flowing during

Opposite. Broad continental shelves in several tropical parts of the world support extensive coral reef systems. This aerial view of the Whitsunday Islands shows part of the largest coral reef in the world, Australia's Great Barrier Reef, almost 2,000 kilometers (1,250 miles) long.

Some continental margins are dissected by steep-walled canyons, scoured out by turbidity currents as they carry sediments to the ocean basins. The longest submarine canyon in the world is the Bering Canyon, north of the Aleutian Islands. The canyon emerges from the continental slope, and makes its sinuous way through the continental margin for more than 1,100 kilometers (680 miles).

times of lowered sea level. The Hudson Canyon off the eastern coast of America is a good example. It is fairly easy to project the present mouth of the river at New York across the shelf to the head of the canyon. The same is true of the huge Monterey Canyon, off the coast of central California. This canyon has depths and a cross-section that compare with the Grand Canyon of Arizona. Although no river enters the ocean today at the canyon's head near Monterey, geological data indicate that the Salinas River entered the sea there during the ice ages, modifying and exhuming this ancient canyon several times.

Some canyons have no river associations, either past or present. Two such are the Bering Canyon, the longest in the world, and the Great Bahama Canyon, with the highest walls of any we know. The Bering Canyon lies on the north side of the Aleutian Islands. It is more than 1,100 kilometers (680 miles) long, has a large number of tributaries, and its head is some 500 kilometers (300 miles) from mainland Alaska. Perhaps the Yukon and Kuskokwim rivers flowed into the Bering Canyon during times of lower sea level, but there is no evidence of this.

The Great Bahama Canyon is indeed a puzzle. It lies between the low Great Abaco and Eleuthera islands yet has wall heights approaching 4,350 meters (14,270 feet): greater than any land canyon! It has a total length of 280 kilometers (175 miles) and two V-shaped branches, one of which heads into the broad floor of the Tongue of the Ocean which itself can be traced southward for 100 kilometers (60 miles) at depths to 1,460 meters (4,800 feet). ■

THE BERING CANYON

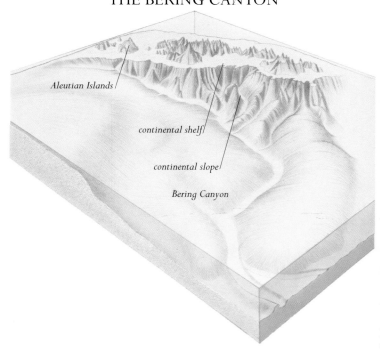

Aleutian Islands

continental shelf

continental slope

Bering Canyon

2 WAVES, TIDES, AND CURRENTS

PAUL SCULLY-POWER

The oceans act as giant heat reservoirs that capture and then slowly release the radiated energy from the sun. They are the most important controlling factor in the world's long-term weather patterns, or climate. This huge global thermostat strongly moderates the range of temperatures and seasonal fluctuations, thereby making it easier for many forms of life to survive. Even small changes in the temperature of the oceans, or their currents, or even in their height, can affect life and property on the remaining one-third of the Earth's surface, the land.

Wind, tides, and currents all contribute to ocean-surface turbulence, like that revealed in this view of the Caribbean Sea from space. The interaction of two major spiral eddies has produced a number of lesser lines of surface turbulence, visible as curved white streaks.

AN OCEAN IN MOTION

Whether you go to sea in a small boat on the Sea of Galilee as recounted in the Bible or travel across the north Atlantic in a luxury liner, you are forever aware that the ocean is in motion. From the lightest breeze through to the wildest gale, the ocean's surface changes accordingly, from a gentle ballet of ruffles to an awe-inspiring and sometimes frightening vista of waves. These wind-created waves, known to mariners as the "sea", are measured by a scale called the "sea state" and logged in numeric values from zero through nine, depending on the wave height: zero sea state corresponds to a calm mirror-like surface, while sea state nine signifies waves over 14 meters (45 feet) in height. In addition, distant storms telegraph their presence by another, more regular, set of waves called the "swell". All these motions can produce the feared *mal de mer* or sea-sickness which is not unlike, but is certainly different from, both air-sickness and space-sickness.

And if this is not enough, the traveler, having reached the safe haven of a port, can go ashore to celebrate, only to return to find that his or her ship is not where it was left; it can be as much as 12 meters (40 feet) higher or lower. This confusing syndrome is not brought about by the onshore celebration, but is a manifestation of the tide.

All these kinds of oceanic movement fall into the general category of waves, which are measured by their height (amplitude), frequency (period), and direction. Added to these are the huge rivers of the ocean, the major current systems, which transport water horizontally over great distances. And finally there is the lesser, vertical, motion of water throughout the volume of the ocean.

SURFACE WAVES

As anyone who has been to sea will surely testify, almost as soon as the wind begins to blow the ocean responds by creating waves; and as the winds continue to blow these waves build up to such an extent that the ship will roll and pitch and heave. The extent of this build-up is governed by three factors: the speed of the wind; the length of time it has been blowing; and the fetch or distance over which it has traveled. The larger any of these three factors, the higher the waves that will be generated. Ultimately, however, an equilibrium is reached where, for a given wind speed, the ocean's surface can absorb no more energy and the waves break. This situation is called a fully developed sea.

Oceanic "swell" refers to surface waves that propagate beyond both the region where they were generated and the local influence of the wind. In traveling a considerable distance these waves undergo an internal ordering so that they become far more regular in appearance and direction. This sorting process results from the spreading of the waves from the generating area under the influence of a phenomenon called dispersion, which causes the longer waves to travel faster than the shorter ones. Hence it is possible to find waves at some distance from a storm, where there may be no wind at all. But these waves are usually relatively smooth with rather long periods, in contrast to the confused waves with different periods that are typical within the area of the storm. Thus the direction of the swell can point to a distant storm.

The majestic roll of breakers on a beach highlights one of the dominant features of the ocean: its restless surface, constantly affected in a multitude of interrelated ways by every wind that blows over it. Someday it may be possible to harness this abundant power to generate electricity, but attempts so far have been largely thwarted by complex engineering difficulties.

Ocean currents result partly from the action of wind on the sea surface, and partly from convection processes deep within the ocean. The world's surface currents together make up five major gyres or broadly circular systems: the North Atlantic, South Atlantic, South Indian, North Pacific, and South Pacific.

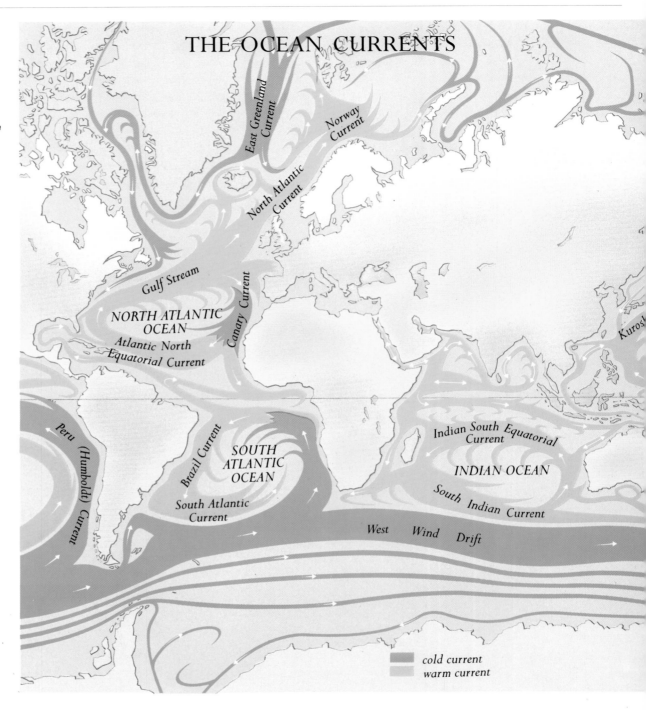

THE OCEAN CURRENTS

East Greenland Current

Norway Current

North Atlantic Current

Gulf Stream

NORTH ATLANTIC OCEAN

Canary Current

Atlantic North Equatorial Current

Kurosh

Peru (Humboldt) Current

Brazil Current

SOUTH ATLANTIC OCEAN

South Atlantic Current

Indian South Equatorial Current

INDIAN OCEAN

South Indian Current

West Wind Drift

cold current
warm current

THE PULSE OF THE OCEAN

Tides are the pulse of the ocean. Their regular rhythm is caused by that most universal of all forces, gravity, and is due to the presence of both the moon and the sun. The force of gravity controls the motion of the planets in orbit, but it is the small differences in the gravitational attraction of the moon (and also of the sun) from one side of the Earth to the other that give rise to the tide-producing forces.

These differences result from the varying distances of points on the Earth's surface from the moon (and the sun). The net result is a slight bulging of the ocean both directly in line with the moon (and the sun) and

also on the diametrically opposite side of the Earth. This twin effect is a result of the difference in gravity being positive on the "near" side and negative on the "far" side; hence water on each side bulges outward relative to the orbital path of the Earth. The combination of these bulging motions, with the sun contributing about half that of the moon, together with the rotation of the Earth about its own axis, produces the typical semidiurnal or half-daily tides with which we are familiar.

It should be noted that tides have nothing to do with tidal waves. These are a misnomer, and really refer to the waves set up in the ocean by underwater

ARCTIC
OCEAN

North Pacific Current

California Current

RTH PACIFIC OCEAN

ic North Equatorial Current

Pacific South Equatorial Current

SOUTH PACIFIC OCEAN

sman
rent

South Pacific Current

earthquakes: waves that can travel undiminished for thousands of kilometers across an ocean basin and which sometimes cause massive damage when they come ashore on a distant coastline. To differentiate, tidal waves are now more commonly called tsunamis, the Japanese term for "tidal wave".

RIVERS IN THE OCEAN

Navigators have known for thousands of years that the ocean is characterized by variable winds and variable currents. However, it is only in the last half century that a reasonably cogent picture has emerged of the patterns of oceanic currents and their underlying causes.

The major force setting up and maintaining the oceanic current system is that of the prevailing winds, and the direction of current flow is governed by these winds and the fact that the Earth is itself rotating. However the way in which currents manifest themselves is quite subtle. Surface winds, especially over the oceans, tend to fall into a regular pattern: near the equator are the doldrums, followed, as you move north or south from the equator, by the trade winds, then the westerlies, and finally the polar easterlies. Basically these are zonal (east–west) bands of wind that alternate in direction: the trade winds blow essentially from the east in the band from the doldrums to about 30° of latitude, the westerlies from there to about 60°, and the polar easterlies from 60° of latitude to the poles.

The effect of these three bands of winds in each hemisphere is to create three regimes of oceanic currents: the equatorial current system, the subtropical gyre, and the subpolar gyre. However, these current regions are not superimposed on the wind regions as you would at first expect, since the ocean does not respond directly to the stress of the wind but rather to the changes in the average winds with latitude. This is because a uniform zonal wind over an ocean would not produce a corresponding current, but

High tide and low tide on a coral beach in Fiji. The height between tidal extremes varies from place to place around the world. It may be almost negligible, as in the Mediterranean, or over 14 meters (46 feet), as in the Bay of Fundy, Canada.

Richard Chesher/Seaphot Limited: Planet Earth Pictures
Richard Chesher/Seaphot Limited: Planet Earth Pictures

would only cause water to pile up on the opposite side of the ocean basin. This results in the rather interesting phenomenon that the major dividing lines between the current systems are not at the latitudes where the winds are zero, but rather at the latitudes of maximum zonal winds. Hence the regimes of oceanic currents are bounded by the latitude bands of 0°–15° for the equatorial current system, 15°–45° for the subtropical gyre, and above 45° for the subpolar gyre.

Added to this, the currents are flowing on a rotating Earth. This introduces another force, called Coriolis, which has the effect of deflecting the currents to the right in the Northern Hemisphere and to the left in the Southern Hemisphere.

GYRES AND ROTATING FORCES

The consequence of the winds being zonally banded and hence changing with latitude, together with the rotation of the Earth, is twofold: firstly, that currents are indeed induced in the underlying ocean; and secondly, that these currents tend to join together and recirculate within each band, thus forming gyres (circular or spiral formations) across their domain of an ocean basin. Hence in the equatorial region, there is a gyre made up of the westerly-flowing equatorial current centered at about 15° latitude and a countercurrent flowing in the opposite (easterly) direction near the equator.

Poleward, to the north and south, the subtropical gyre circulates in the opposite direction, being formed by the westerly equatorial current at 15° latitude and the easterly-flowing "west wind drift" at about 45° latitude. Further poleward, the circulation is again in the opposite direction, at least in the Northern Hemisphere where the ocean basins are closed off by the landmasses. In the Southern Hemisphere, the ocean basins are not similarly closed off, and the "west wind drifts" in each ocean join together to form the

WAVE FORMATION

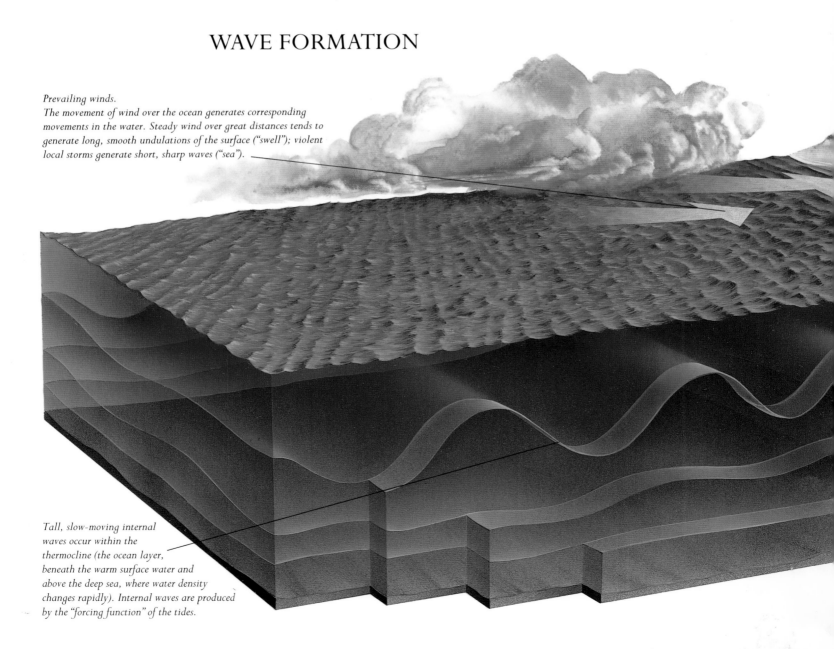

Prevailing winds.
The movement of wind over the ocean generates corresponding movements in the water. Steady wind over great distances tends to generate long, smooth undulations of the surface ("swell"); violent local storms generate short, sharp waves ("sea").

Tall, slow-moving internal waves occur within the thermocline (the ocean layer, beneath the warm surface water and above the deep sea, where water density changes rapidly). Internal waves are produced by the "forcing function" of the tides.

Antarctic Circumpolar Current, the only current in the world that completely circumnavigates the globe.

There is an additional, but most important, aspect of the Earth's rotation, which is that the amount of local rotation changes with latitude. To understand this, imagine yourself standing at the North Pole, in which case you would be rotating in a counter-clockwise direction; if you were standing at the South Pole, however, you would be rotating in a clockwise direction. The local rotation changes direction at the equator, and indeed is zero there. This change in the direction of rotation causes the currents in the gyres to flow in opposite directions on each side of the equator.

Furthermore, the increasing local value of the rotation as you move from the equator to the poles causes the currents on the western sides of ocean basins to be much stronger than those on the eastern sides. This "western intensification" explains the existence of such mighty currents as the Gulf Stream off the east coast of North America and the Kuroshio off the east coast of Japan. An appreciation of the magnitude of these currents can be obtained by considering that the Gulf Stream has current speeds ranging from about half a knot to more than 3 knots, and carries about 135 billion liters (30 billion gallons) of water every second, which is about 65 times that of all the rivers in the world. In this sense, therefore, these "rivers of the ocean" are far more majestic than their land counterparts.

RINGS AND SPIRAL EDDIES

The fact that these currents are so strong often leads to another phenomenon, that of oceanic eddies. When the speed and intensity of, for example, the Gulf Stream reaches a critical value the current itself becomes unstable, oscillates in direction, and spins off an eddy about 160 kilometers (100 miles) in diameter by pinching off this large oscillation or meander. Because

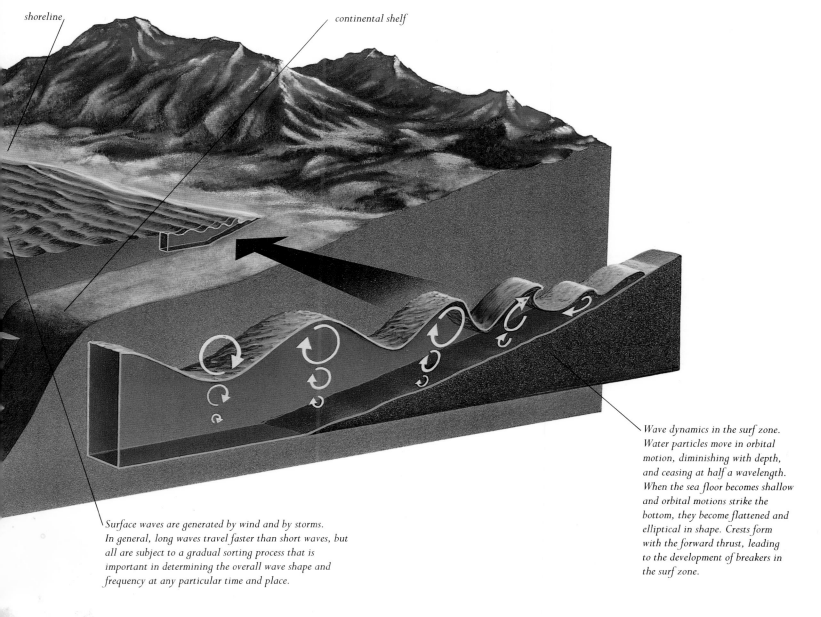

shoreline

continental shelf

Surface waves are generated by wind and by storms. In general, long waves travel faster than short waves, but all are subject to a gradual sorting process that is important in determining the overall wave shape and frequency at any particular time and place.

Wave dynamics in the surf zone. Water particles move in orbital motion, diminishing with depth, and ceasing at half a wavelength. When the sea floor becomes shallow and orbital motions strike the bottom, they become flattened and elliptical in shape. Crests form with the forward thrust, leading to the development of breakers in the surf zone.

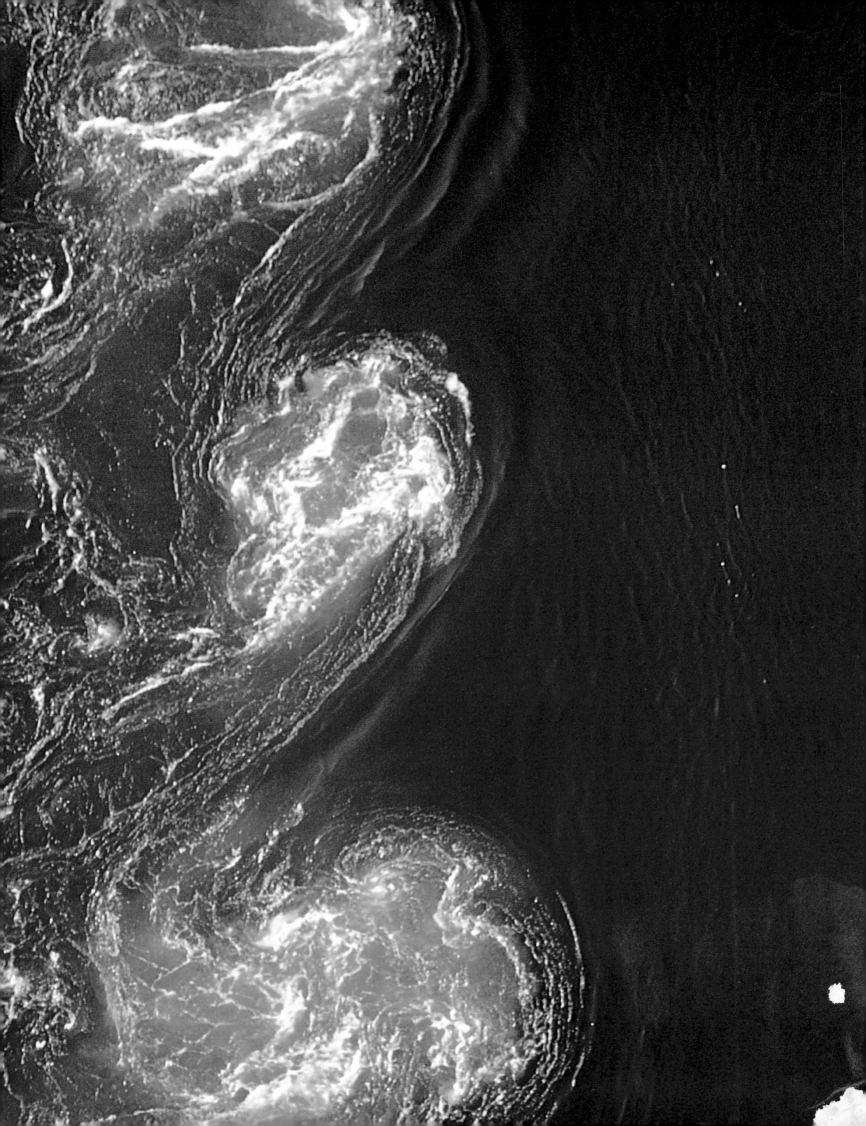

of their near-circular shape these eddies are sometimes called rings. They often move in the opposite direction to the main current that generated them, and can retain their identity for up to three years. Perhaps this is nature's way of maintaining stability, by first limiting energy build-up and then redistributing that energy.

Recently, through observations made from space, another type of eddy has been discovered. These have a spiral shape rather than the near-circular shape of rings, and are considerably smaller, being only about 16 kilometers (10 miles) in diameter. Little is yet known of these spiral eddies, other than that they all have a similar structure and are being discovered increasingly in many areas of the ocean. They appear to be made up of highly sheared currents: the strength of the current changes in steps as you move out from the center of the eddy. Again, it is perhaps nature's way of maintaining equilibrium.

VERTICAL MOVEMENT

The oceans also support vertical motion to a limited but vital extent. To understand the magnitude of this, one must first consider the disparity between the scales involved. Ocean basins are typically several thousand kilometers wide, but are only several kilometers deep even at their deepest points; hence the vertical motions should be reduced by at least this same factor.

Added to this, the oceans are horizontally stratified, which means that they are layered horizontally. As you go deeper and deeper in the ocean, the temperature progressively drops from around 25°C (77°F) near the surface at low latitudes to close to 0°C (32°F) toward the bottom. Because the oceans are heated by the sun from the top down, they form horizontal layers with the warmer and hence less dense water near the surface, and colder and denser water further down. Since this is a very stable situation, it is unlikely that this stability will be upset by vertical motion.

These combined effects severely limit the magnitude of vertical movement in the ocean. However, since the surface waters near the poles are themselves very cold, there does exist an extremely slow flow of this cold, dense water outward from the poles toward the equator, progressively sinking as it moves to its equilibrium depth. Over periods of hundreds of years this gradually replenishes the deep water throughout the oceans.

The most significant consequence of this vertical ordering in the oceans is that the horizontal currents, even the strongest ones like the Gulf Stream and the Kuroshio, are significantly reduced with increasing depth, to such an extent that they are virtually non-existent at depths of around 1,500 meters (5,000 feet). If there were no vertical motion, the horizontal

THE FORMATION OF A GULF STREAM RING

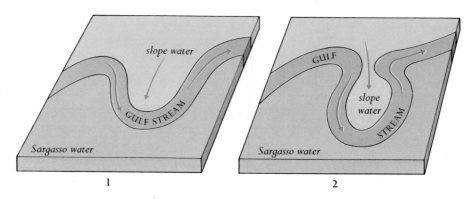

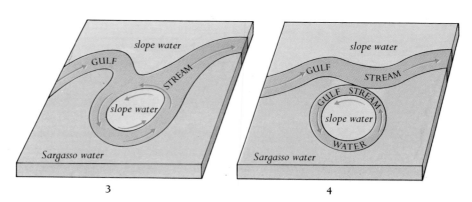

currents would extend from the top to the bottom of the ocean and remain undiminished in strength.

The only regions where this situation does not apply is where local winds blow the waters away from the coastline, and deeper waters must then flow vertically upward to fill the void. This is particularly evident along the eastern sides of ocean basins (the western sides of continents) where the local winds have a significant meridional, or north–south, component. Then, because of the effect of Coriolis, the water is transported away from the shore, and deeper, colder, and nutrient-laden water is brought up from below. These "upwelling regions" are important for the fishing industries because the upwelling enriches the plankton that is the basis of the oceanic food chain.

THE WAY AHEAD

The dynamics of the oceans present a tapestry of complex patterns which, given their diversity of size and persistence, make them difficult to monitor on a continuing basis.

Fortunately, two modern technologies are now at the stage where they can combine to solve the problem: space remote sensing and ocean acoustic tomography, the latter being the equivalent of a CAT scan of the ocean using acoustic energy. On the one hand, space sensing allows the surface of the ocean to be measured regularly; while on the other acoustic tomography can tell us much about the interior of the ocean. ∎

The Gulf Stream, the strongest current in the North Atlantic Gyre, is characterized by meanders, eddies, and self-contained and sometimes long-lived rings, which break off from the main current as a result of their speed and intensity.

Opposite. At several places around the world, such as here in the Saltstraumen channel off the northwestern coast of Norway, the distribution of land creates a collision of strong currents and tides. Here there are whirlpools and dangerous rapids in the sea, and the thundering roar of contending waters can be heard up to 5 kilometers (3 miles) away.

3 THE FRINGES OF THE SEA

ALEC C. BROWN, M.G. CHAPMAN, AND A.J. UNDERWOOD

The meeting of land and sea produces a range of environments where plants and animals thrive on sandy beaches and rocky shores, in river estuaries, dense mangrove forests, saltmarshes, and seagrass meadows. The habitat of the plants and animals of this zone is sometimes harsh and always unstable, often buffeted by waves and subject to the constant ebbing and flowing of tides. The organisms that inhabit the fringes of the sea have adapted to their unpredictable environment in a remarkable variety of ways.

Sandy beaches are home to a range of microscopic creatures. The illustration indicates a representative sample:

A *mystacocoid* Derocheilocaris
B *copepod* Hastigerella
C *oligochaete* Marionina
D *nematode* Nannolaimus
E *gastrotrich* Xenotrichula
F *archiannelid* Poligordius
G *tubellarian* Diascorhynchus

SANDY SHORES

Sandy shores make up some 75 percent of the world's ice-free coastlines. The occurrence and structure of these extremely dynamic environments depend on movements of air, water, and sand. Patterns of water movement determine whether a stretch of coast will be predominantly eroding or depositing. In the latter case the force of the waves determines the size of the particles deposited: minimal wave action is likely to result in a muddy shore, greater wave energy in a sandy beach.

Most sand consists mainly of quartz (or silica) which originates from inland erosion and reaches the sea by way of river systems. Some sand, however, has its origin in the skeletons of marine animals, the shells of mollusks, or cliff erosion. This sand consists predominantly of calcium carbonate. Other materials which may contribute to beach sands include heavy minerals, basalt, and feldspar.

There are many kinds of beaches. The extremes are *reflective*, in which the beach slopes steeply both intertidally and offshore; and *dissipative*, where the beach shelves gently, with a system of bars and shallow troughs offshore. On reflective beaches, waves break on the beach itself and the wave energy is reflected back to sea. On fully dissipative beaches, however, the waves break on bars or sandbanks well out from the beach, so that their energy is largely dissipated before they reach the intertidal zone. The waves then give rise to a relatively gentle swash running up and down the beach. Dissipative beaches thus have broad surf zones and are of much greater recreational value than reflective beaches.

The width of a beach, and to a large extent its slope, are determined by cycles of deposition and erosion. In those few places where deposition consistently exceeds erosion, the beach marches steadily out to sea. For example, Hastings in southern England, which was on the sea when the Normans invaded in 1066 (and for a long time after then), now lies several kilometers inland. It is more usual, however, either for cycles of deposition and erosion to balance each other or for erosion slightly to exceed deposition, so that the beach slowly retreats. Most erosion occurs during storms, when sand is removed from the intertidal zone and dumped in the surf zone. It slowly returns to the beach as conditions return to normal. In this way, the beach and its surf zone act as a wave-dissipating apron, absorbing wave energy through the movement of sand and thus protecting the land behind.

Wave action, the size of the sand particles, and beach slope are all interrelated. In general, the greater the wave action, the greater the slope and the greater the size of the sand particles. However, this is complicated by the fact that cycles of deposition and erosion are more extreme on exposed than on sheltered beaches, so that exposed beaches show greater variation in slope over a period of time. Cusps, or mounds of sediment, are a common feature of sandy beaches and result from the interaction of bars, troughs, wave action, and currents.

BEACHES IN MOTION

Sandy beaches act as giant filters. Much of the water in the swashes running up the beach percolates down through the sand until it reaches the water table. This creates a hydrostatic pressure which forces water out at

A beach undergoes constant flux, its form dependent on the availability of sediment and the force of the waves. On shallow sloping beaches such as this one in Namibia, much of the movement of sediment is simply an exchange between offshore bars (evidenced by waves breaking far off the beach) and the dunes (berm) at the top of the beach.

Beach dune systems are often stabilized by salt-tolerant vegetation, such as these strand morning-glories Ipomoea pescaprae *on a beach in Fiji. Common along tropical shores throughout the Indo-Pacific region, this creeper's leaves and flowers grow from a tangle of trailing stems that may extend up to 18 meters (60 feet) in length.*

the bottom of the beach. As a result there is a flow of water through the beach, particularly when the sand is coarse, and particles in the water are trapped between the grains. These particles are largely organic and form a food supply for bacteria, which metabolize them and return enormous quantities of dissolved nutrients to the sea. Sandy beaches have therefore been described as huge "digestive systems".

Sand moves not only up and down the shore but also along the shore, especially in the surf zone, through which large amounts of sand are transported during storms. Although the longshore movement of sand may change direction according to various circumstances, its net annual movement is usually in one direction only, a phenomenon known as "net littoral drift". This creates problems for coastal planners and marine engineers, as any structure which impedes longshore sand movement results in the deposition of sand on its updrift side and erosion downdrift. A classic example is Madras harbor in India, the construction of which resulted in changes to the sandy shoreline for several kilometers as well as severe shoaling of the harbor entrance. As a consequence, the entrance had to be moved and an outer quay constructed.

prawns are common on sheltered beaches, while on exposed beaches smaller forms, such as sand lice (Isopoda), are more likely to predominate. On tropical beaches, mole crabs *Emerita* often occur in considerable numbers. Crustacea are also common at the top of the shore, around the driftline, where they have adapted to a semi-terrestrial existence. In addition, sand hoppers are important here, particularly in temperate climes, whereas in the tropics the driftline is quite often dominated by ghost crabs *Ocypode*.

The surf zone supports a rich fauna and acts as a nursery area for many fish. Among the invertebrate surf-zone fauna, swimming prawns and shrimps predominate, together with smaller crustaceans such as copepods. The surf-zone fauna is an integral part of the sandy beach ecosystem, because these animals invade the intertidal zone as the tide rises and many are predators on intertidal invertebrates. Just as the beach is invaded by surf-zone animals with the rising tide, so it becomes available to terrestrial forms during ebb. Many birds feed on the intertidal fauna during the day, and some even at night, when they may be joined by a number of other predators and scavengers.

HOW SANDY BEACHES ACT AS FILTERS

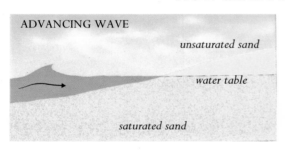

ADVANCING WAVE

unsaturated sand

water table

saturated sand

HIGH WAVE

some water percolates through sand

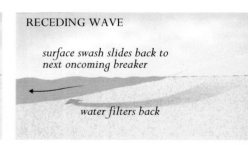

RECEDING WAVE

surface swash slides back to next oncoming breaker

water filters back

As waves constantly break on the beach, water percolates through the sand, ensuring a constant flow of nutrients which filter through the beach.

SANDY BEACH FAUNA

Although sandy beaches teem with animal life, most species are only just visible to the naked eye. They live between the sand grains, never willingly leaving their secure habitat, so are rarely seen by the casual observer. Most animal phyla are represented in this minuscule fauna and some groups are found nowhere else on Earth. Although small, most of these animals are very mobile, migrating up and down through the sand with the tides or in response to factors such as temperature or light.

There are fewer species of relatively large animals, although individual species become numerous under suitable conditions. Polychaete worms, such as lugworms and bloodworms, are common, especially on relatively sheltered beaches, while clams and whelks can attain vast populations on more exposed shores. Crustacea are also important: burrowing

ON THE MOVE

A characteristic of all intertidal sandy beach invertebrates is their ability to burrow into the sand. This protects them not only from predators but also from desiccation and extremes of temperature. Some adjust their depth to prevailing conditions, burrowing deeper during cold or hot weather, or when storms threaten to wash them out of the sand.

Burrowing is often accomplished with surprising speed: the mole crab can bury itself completely in less than 1.5 seconds. At the other end of the scale, the lugworm *Arenicola* may take several minutes. There is some correlation between speed of burrowing and exposure to wave action. Under exposed conditions, if it is not to be swept away by the next surge of water, a burrowing animal must be able to obtain a firm anchorage in the sand between swashes. *Arenicola* can thus burrow only in quiet water; however this is consistent with the fact that it lives in a semi-permanent burrow which it can keep open only where wave action is slight. On beaches more exposed to wave action, the sand is too unstable to support burrows, and the animals must be more mobile and agile if they are to maintain their position on the beach.

Most of the larger species migrate up and down the beach with the tides, often surfing in the waves. By doing this, the animals maximize their food supply, which is often most abundant at the water's edge, and

thus deprived both of shelter and of a resident primary food source, and must rely on food imported from the land or sea. Although debris and insects may be blown in from the dunes, and the bodies of birds and other terrestrial animals may end up on the beach, nearly all the food of intertidal fauna is of marine origin.

There are three intertidal food source systems: one based on phytoplankton washing up from the surf zone; one based on an input of kelp or wrack; and one dependent on carrion such as stranded jellyfish. These

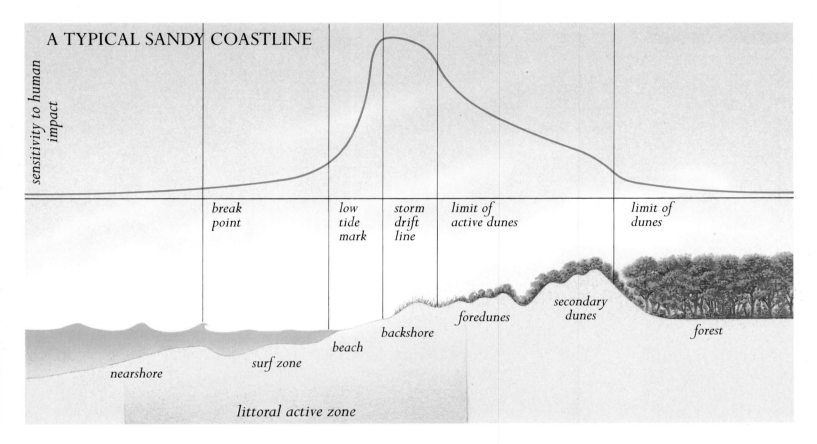

A TYPICAL SANDY COASTLINE

sensitivity to human impact

break point

low tide mark

storm drift line

limit of active dunes

limit of dunes

foredunes

secondary dunes

forest

backshore

beach

surf zone

nearshore

littoral active zone

also make it difficult for predators to reach them. An example of this behavior is provided by the beach clam *Donax,* which surfs up the shore with its large foot and siphons extended to take full advantage of the waves. Arriving in the swash zone, it buries itself rapidly and uses its siphons to filter feed, emerging as the tide rises to surf once more in the swash zone. It avoids the upper part of the shore, where it might be stranded, and surfs down the slope again as the tide falls. Whelks such as *Terebra* and the plough snail *Bullia* also surf by using the fully expanded foot as an underwater sail, while the mole crab allows the waves to transport it as a tightly rolled ball.

SOURCES OF FOOD

Sandy beaches are an extremely harsh habitat, not only because of wave action and sand movements, but also because the instability of the substratum precludes colonization by attached plants. Intertidal animals are

systems are not mutually exclusive but the major food source determines to a large extent the type of animal that can colonize the beach. Where the food consists mainly of phytoplankton, filter feeders such as *Donax* or *Emerita* dominate the beach; where carrion is the chief input, scavengers like *Bullia* and crabs come into their own; if kelp or wrack is the main food source, the center of gravity of the community moves toward the top of the shore and the animals consist predominantly of semi-terrestrial herbivores—for example, crustaceans such as talitrid amphipods and large isopods.

The survival of sandy beach animals depends on one thing—adaptability; the ability to adapt to changing conditions of wave action, to cycles of deposition and erosion, to changing temperatures, and above all to an erratic and varied food supply. They have little room for specialization and, like us, are born opportunists.

ALEC C. BROWN

A cross-section of a typical sandy coastline. The sensitivity curve indicates the areas most sensitive to human impact.

ROCKY SHORES

There are several kinds of intertidal rocky shores: vertical cliffs, sloping shores, platforms cut by the action of waves, and areas that are littered with boulders. These varied shores often support different sets of plants and animals, nearly all of which are fully marine in origin. Few terrestrial species are adapted for life between the tides on rocky shores.

THE IMPORTANCE OF WATER

Two major physical influences act on the organisms in rocky intertidal habitats. First are the periodic and predictable patterns of the tidal rise and fall of water. In most areas of the world, the tide rises and falls roughly every 13 hours. As a result, marine plants and animals at the top of the shore are out of the water for a relatively long period before the tide rises again. They must therefore withstand periods of increased or decreased air temperature and desiccation during their time out of water. Organisms lower down the shore have to withstand only short periods out of water.

Coupled with this pattern is the regular fortnightly progression from neap to spring tides and back again. The water rises to higher levels and falls to lower levels during spring tides than during neap tides. During neap tides, organisms high on the shore may not be covered by water for several days; at the bottom of the shore, they may be under water continuously. In either case, prolonged exposure to one condition may impose severe hardships.

The second major physical factor is the force of waves. On wave-exposed shores, the effects of severe wave-shock are such that only flexible plants that are not damaged when hit by waves, or tough animals, such as limpets and barnacles that are well stuck down, are able to survive. Wave-exposed shores are often covered by spray, thus reducing the possible harmful effects to organisms of being out of water during low tide.

The effects of these processes vary according to the type of rocky shore. Vertical cliffs offer no shelter from waves, and plants and animals at all heights are subject to extreme wave-force on exposed shores. In contrast, on boulder-fields there are numerous spaces between the boulders and underneath the rocks that provide shelter from the impact of the waves.

LIFE BETWEEN THE TIDES

Plants and animals living at the top of the shore must be resistant to long periods of drying out during low tide. The plants at these high levels are able to survive even when they are dried to a crisp at low water. When the tide eventually comes in, these seaweeds rehydrate and begin to function normally.

The most common animals of the high shore are small grazing snails, which have a number of characteristics that

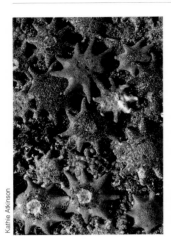

The multi-hued eight-rayed starfish Patiriella calcar *is usually found low on the shore or in rockpools.*

Kathie Atkinson

The periwinkle moves on a muscular, fleshy foot, like that of the land snail. Its shell is thick and rounded, which allows it to roll into a crevice should it become detached.

Though they cling like limpets, barnacles are actually crustaceans, not mollusks. At high tide the barnacles open their shell-like carapace and extend their feathery limbs (called cirri) to feed, raking in tiny food particles floating by. The shell is sealed at low tide to prevent drying out.

Common in rockpools almost everywhere, hermit crabs are soft-bodied, and live permanently in the abandoned shells of mollusks. Carrying the shell about with them wherever they go, the crabs must periodically trade up to larger shells to keep pace with their growth.

A rockpool on the Californian coast. The overall structure of the animal community inhabiting tidal rockpools differs little throughout the temperate regions of the world, though the species themselves may vary. The inhabitants usually have at least one thing in common: they must routinely cope with the daily tidal cycle of flood alternating with exposure to sun and air.

LIFE IN A TIDAL ROCKPOOL

Members of the bivalve group of mollusks, mussels often grow in dense clumps or colonies. Some species, like the edible mussel Mytilus edulis, are found virtually worldwide.

Seaweeds are marine algae; they differ from land plants in lacking roots, flowers, seeds, or fruit. Seaweeds anchor themselves in place by means of a sucker-like structure called a holdfast.

Small octopuses are common inhabitants of rockpools, but generally they are extremely shy and elusive, and seldom seen.

Although some sculpin species live in deep water and others in fresh water, many are common and characteristic inhabitants of rockpools around the world. They are predators that prefer to lie in wait for their prey rather than chase it.

Blennies spend a great deal of their time hiding under stones, or darting from one shadowed cavity to the next. They lack the swim bladder common to most fishes of the open sea.

Often brightly colored, sea-anemones are animals, not plants, that feed on smaller animals that have been subdued by stinging threads in the sea-anemones' waving tentacles. When not feeding, they retract their tentacles and close up into a jelly-like blob.

Small crabs are abundant in rockpools almost everywhere, sometimes with several species occurring together.

Starfish are carnivorous animals that move slowly from place to place on hundreds of tiny feet which operate on water pressure. They turn their stomachs inside out to engulf their prey, which is usually mollusks of various kinds.

prevent them from becoming dried out. Often they remain immobile when the tide is further down the shore, moving to feed only when the tides are high. Many also move around and feed when it rains. When immobile, they commonly withdraw into their shells, which are glued to the rocks by a ring of dried mucus or slime.

At the bottom of the shore, plants and animals are larger, there are more species, and they must cope with different problems. Many species, such as large barnacles and sea-squirts, are sessile—they are stuck to the rocks and cannot move. Others, such as branching seaweeds, sponges, and sea-anemones, are soft-bodied. Rarely do these plants and animals need to survive prolonged periods of drying out. Even during spring low tides, splash and spray often drift over the lower parts of the shore, except during the calmest weather, and even then the air is usually humid because of the closeness of the sea. Filter-feeding animals (barnacles, sponges, mussels, and sea-squirts) predominate as they can feed for long periods when the tide is in.

Toward the bottom of the shore, competitive overgrowth of some seaweeds by others, and swamping of barnacles or mussels by seaweed, are common. As a result, some species are reduced in numbers or coverage of the shore.

Low on the shore, plants and animals are subject to predation or grazing by animals coming upshore with the tide. Biological processes of competition and predation are much more important than physical factors as controls on the types, numbers, and sizes of plants and animals at low levels.

A PATCHY HABITAT

Rocky shores on many of the world's coastlines are patchy habitats—a feature caused by a variety of processes. Patterns of grazing or predation are linked with features of topography such as holes or crevices in the rocks. Often, grazing or predatory mollusks shelter in cracks or crevices to avoid predators such as crabs or fish, or to reduce the intensity of physical stress (cracks and crevices are shaded and damp). Sheltering animals emerge at high tide to feed on surrounding organisms, but cannot move far from shelter because they must get back into a crevice before the tide falls or before the next period of inclement weather. Thus, when suitable shelters are sufficiently far apart, foraging causes patchy reductions in numbers, or even the disappearance, of prey around these shelters. Such "haloes" are widespread on rocky shores.

Opposite patches are formed when prey species manage to grow in, or move to, areas that are inaccessible to their consumers. Thus, some species of algae thrive in the middle of large patches of barnacles because their grazers do not have enough room among the barnacles to reach them.

Predation by large predators that come upshore with the tide can also be a direct cause of patchiness. As the tide rises, some starfish, fish, and crabs move into inter-tidal areas to feed on seaweeds or animals. In Chile, for

The Photo Library-Sydney/Geoff Higgins

The ceaseless pounding of the sea unites with slow wind-abrasion in shaping rocky coastlines. In this tessellated pavement at Eaglehawk Neck, Tasmania, strata of soft sandstone has eroded somewhat faster than an overlying layer of harder sandstone. The resulting pools and crevices harbor rich animal communities.

Opposite. *Dense thickets and underwater forests of bull kelp* Durvillea potatorum *fringe the littoral zone (the region between the tides) in many parts of southeastern Australia, New Zealand, and on many subantarctic islands.*

Many small animals of the tidal zone have evolved various clinging mechanisms both as a defense against predators and as protection from being bumped and battered about by the surf. As seen from underneath, seastars (center) have tiny suckers at the tips of hundreds of tiny tubed feet, which cling by water pressure rather than muscle power. Others, such as limpets (top) and periwinkles (bottom), cling by means of a large muscular foot.

example, there is a clingfish that arrives with the waves at mid-to-high levels on the shore. It attaches its sucker to the rocks and rasps off animals and plants from the shore before releasing itself and returning to the water.

Another major cause of patchiness is physical disturbance by waves, particularly during storms. For example, if waves overturn a boulder on a boulder-field, many of the plants and animals on the boulder will die. Some will be scraped off as the boulder moves against the rocks. Moreover, the organisms on the underside are now out in the open and those on the top are now in the dark: both are in a different habitat and may die as a result, or because predators can now reach them.

Thus, new cleared surfaces become available for colonization by other plants and animals. If the boulder is not disturbed again for a long time, there will usually be competition for space among the organisms on each surface, leading to the eventual re-establishment of a uniform, non-patchy appearance throughout the boulder-field. Disturbances by waves affect different boulders at different times and boulders of various sizes at different rates. The result is a mosaic of patches—boulders at various stages of development, from newly overturned to completely undisturbed.

COLONIZATION OF THE SHORELINE

The primary cause of patchiness in the numbers and species of plants and animals on rocky shores is the process of colonization. Most intertidal species have a life cycle that includes a dispersive larval stage. These organisms reproduce by shedding their spores or larvae into the sea, where they develop—often through a complex series of larval stages—before being washed back into an intertidal habitat. During this period of development, the planktonic offspring may be washed many kilometers along the coastline by winds, currents, and tides. Large numbers are eaten by predators, ranging from other planktonic larvae to fish. Usually only a small percentage survives, and these are presumably widely scattered. When their development is complete, they can colonize a suitable empty patch of shore. The animals hastily attach themselves to the rocks and metamorphose to take on the adult body form.

The plants and animals that can colonize a patch will therefore depend upon which larvae and spores are in the water when the patch is cleared. This can result in different mixtures of species in different numbers, from one place or time to another, thus maintaining the patchy appearance of the shore.

Rocky shores are characterized by dynamic, variable processes of recruitment, competition, predation, disturbance, and physical stress. These vary in rate and intensity from one height on a shore to another, along a gradient of exposure to wave-force, and from place to place and time to time. The result is a diversity of interactions among the species and, usually, a changing structure that provides endless fascination for the observer.

THREE STAGES IN THE LIFE OF A CRAB

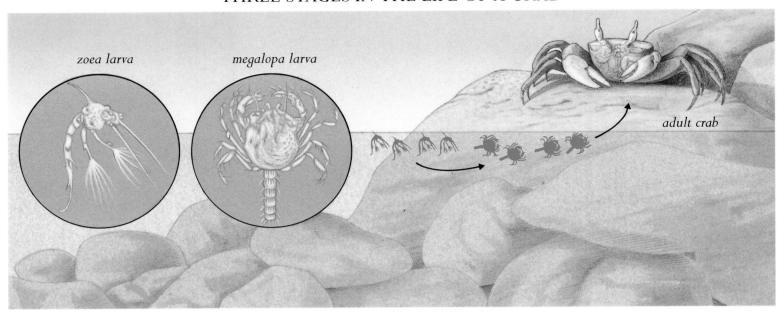

zoea larva

megalopa larva

adult crab

Crabs, like most intertidal animals, reproduce by releasing their larvae into the sea, where they develop through several larval forms before returning to an intertidal habitat. Larvae may be dispersed over considerable distances, and only a small percentage of them survive to colonize a rocky shore.

Opposite. Low tide exposes the magnificent white sand swirls carved by the ever-changing channels in this estuary in Queensland, Australia. Protected from the fresh water of the estuary, a small fringing reef nestles behind the peninsula.

THE ESTUARINE ENVIRONMENT

Estuaries are the interface between rivers and the ocean, an environment where fresh water from a river runs through a low-lying coastal plain and meets oceanic waters in a bay or channel that is semi-enclosed but has a connection to the open sea. The two types of water mix to produce a habitat with reduced salinity.

Estuaries can be formed through several geological processes. Some, such as the Severn and Thames estuaries in Britain and the mouth of the Amazon River, are drowned river valleys. These are extensive shallow estuaries formed when the sea rose so that sea water entered and swamped the river system. The large amounts of sediment deposited by the rivers form the bottom of the estuary. A second form of estuary occurs where wave action creates a build-up of sand across the mouth of a river, eventually forming a partial barrier, behind which fresh water is trapped and mixed with sea water. An example is the extensive Waddensee in the Netherlands. Fiords, common in Norway, are a third form of estuary, created where a river ends in a deep area of water isolated from the sea by a "sill" or "step" at the oceanic end. The sill is a rocky barrier creating only a shallow opening to the outside sea. Finally, movements of the Earth's crust, during earthquakes or volcanic eruptions, can cause depression of an area of coastline, which sinks to form a bay with a narrow opening to the sea. San Francisco Bay, the Red Sea, and the Gulf of California are well-known examples.

Because their mouths are relatively narrow, estuaries are usually sheltered from the force of waves. They are, however, subject to periods of rising and falling tide. A rising tide brings in oceanic water, making the estuary more salty. When the tide falls, water leaves the estuary and the influx of fresh water from the river reduces salinity. Plants and animals in an estuary therefore live in an environment of cyclic change in the water. There can also be unpredictable changes. After periods of great rainfall, the rivers flow more rapidly and more fresh water enters the estuary. After rain, rivers can bring considerable amounts of silt into an estuary, making the water cloudy or turbid.

MANGROVES AND MANGROVE FORESTS

Mangroves are a feature of many tropical estuaries and can form extensive forests. There is a wide variety of mangrove trees, ranging from small shrubs to large trees. Despite their obvious differences, they have several features in common. They can all tolerate living in soft, waterlogged, often anoxic mud (mud without air), and can cope with a daily inundation of salty water.

Mangroves, like all trees, take up water via their roots. But large amounts of salt are toxic. They cope with the toxicity in different ways. Some prevent the salt from entering their roots with the water by physiological cellular processes. Some absorb the salt and then secrete it out of small pores in their leaves. Others collect the salt in particular parts of the tree, sometimes old leaves, which they then shed.

Mangrove trees have aerial roots to assist with the exchange of gases with the air, which is difficult because the mud is often anoxic and waterlogged. Some roots (stilt-roots) arise from the stem of the plant, well above the ground, and help to support the plant in addition to helping it to breathe. Others (pneumatophores or peg roots) come up through the ground from the shallow horizontally spread root system. Because of their widespread matted root system, mangrove trees trap

Below right. *A giant among crabs, the mudcrab* Scylla serrata *is common in mangroves across much of the Indo-Pacific region. Unlike most other mangrove crabs, which usually emerge at low tide, the mudcrab shelters in its burrow while the tide is out, emerging to forage at high tide. Mainly nocturnal, it is a carnivore and scavenger that feeds on almost anything it can overpower, including individuals of its own species.*

An underwater view of the root system of the mangrove Rhizophora stylosa. One of several devices that mangroves use to secure a stable footing in the shifting ooze in which they grow is to throw multiple roots outward in a stilt-like configuration, a characteristic feature of this particular genus of mangrove.

consolidate silt, creating a stable habitat for many other plants and animals and ensuring an appropriate environment for the trees. Suitable soil is so important that several species of mangrove trees develop their seedlings to an advanced stage before they are released from the parent tree. As a result, seedlings are able to establish themselves quickly before they are washed away from the soil.

Other mangrove trees produce floating seeds which can be washed to a different part of the forest before they sink to the bottom and develop. Mangrove forests provide shady and moist habitats for a diversity of plants and animals. Some move into the mangroves from the land; others come in from the sea, using the forests as an extension of their more usual habitats. Other species are specialized inhabitants of mangrove forests and do not occur anywhere else. Some mistletoes, for example, occur almost exclusively on a few species of mangroves.

The upper branches, leaves, and trunks of mangrove trees are a terrestrial environment. Many spiders, insects, and birds live on the trees, just as they do in

a terrestrial forest. The lower parts of the mangrove trunks and the aerial roots themselves often serve as a home for marine animals such as barnacles and oysters, which pack onto these hard surfaces because they cannot survive on or in the mud. Other marine animals, however, live on the surface of the mud, or burrow into it. Particularly striking in many parts of the world are the vividly colored, busy crabs that move over the mud to feed on algae, but dart back into their burrows when disturbed or to avoid being eaten by predators, which come into the forests with the rising tide.

SALTMARSHES

Saltmarshes are commonly found in estuaries. If there are mangrove forests, the marshes are behind them. If there are no mangrove forests, saltmarshes can form large habitats and serve to consolidate sediments. Typically, they are characterized by grasses and shrubby or prostrate plants rather than large trees. Like mangroves, however, the plants in saltmarshes must cope with salty water regularly arriving with the tide. Interestingly, many saltmarsh plants have succulent

leaves, a feature more commonly associated with arid habitats. Although the plants are surrounded by water for much of the time, the water is saline and it is thus difficult for the plants to take it up into their tissues. It is as if they were living in a very dry habitat.

As in mangrove forests, many species of insects are found in saltmarshes, although most are also widespread in other habitats. There are some animals, such as snails and crabs, that are similar to, or the same as, those that are found in mangroves. Saltmarshes often form important habitats for birds that feed on marsh plants, particularly migratory species such as ducks and geese.

SEAGRASS MEADOWS

Seagrasses are marine plants derived from terrestrial forms, but living mostly under water in estuaries, where they form extensive meadows. Unlike seaweeds, which usually attach to hard rocky surfaces by means of a holdfast (a specialized sucker at the base of the plant), seagrasses have proper roots which anchor them in the mud. They produce flowers and seeds that float to new sites and germinate to develop into new plants. New

vegetative growth also arises from rhizomes—specialized branching stems that run under the sand putting up new shoots. These shoots bear elongated leaves similar to those on grasses, although seagrasses are not related to true grasses.

Seagrasses are associated with an array of other plants and animals because their leaves provide stable, hard surfaces which support seaweeds and small animals that cannot live in soft sediments. Many other animals burrow around the root systems of the seagrasses, which offer protection from predators and make burrowing easier.

In some areas, intertidal regions of estuaries are not stabilized by plants and form continuous mudflats. These often support large numbers of marine or estuarine animals, particularly small crustaceans, worms, and some snails and bivalves that feed on the abundant microalgae growing on and in the mud. These animals, in turn, provide food for many species and wading birds. ■

A.J. UNDERWOOD AND M.G. CHAPMAN

Evaporation rates are high where Australia's torrid Great Sandy Desert meets the sea, resulting in the glittering salt encrustations on this white mangrove Avicennia marina *seedling. Mangroves are characterized by their extremely high tolerance of salt.*

KELP FORESTS

MARGARET ATKINSON

David Doubilet

An underwater forest of kelp waves in the currents and tidal surges off Santa Catalina Island. Commonly called elk kelp, this particular species, Pelagophycus porra, is restricted to California waters.

Kelps, *a form of brown algae, are conspicuous marine plants that usually grow on rocky reefs in temperate water (5–22°C/41–72°F), and are common along stretches of open coast. The best-known species is the giant kelp* Macrocystis pyrifera, *which grows in cold temperate waters of both the Northern and Southern hemispheres. Other familiar kelp plants include* Ecklonia radiata, *from the Southern Hemisphere, and* Laminaria *species from the Northern Hemisphere.*

Although there are several different species of kelp, two general growth forms can be recognized: those that have a simple long trunk (thallus) reaching heights of between 50 centimeters (20 inches) and 2.5 meters (8 feet), with a frond branching from the top of the thallus; and those, like the giant kelp, which grow to extraordinary lengths and have fronds appearing all the way up the thallus. Giant kelp may grow to over 50 meters (165 feet); to scuba divers it looks like a cathedral of enormous plants. Some kelps, including

the vast majority never complete the life cycle. However, those that survive can grow very quickly: giant kelp, for example, often exceeds 30 centimeters (12 inches) per day, which makes it the world's fastest growing plant.

Because kelps often form dense canopies, either on the surface or a meter or so above the sea floor, groups of plants are often referred to as "forests" or "stands". In terms of their interaction with their environment, they do behave rather like forests on land. Not only do they provide shelter and food for many species of mobile animals, including fish, sea-urchins, crabs, and sea-otters, they also reduce light levels to the sea floor, thereby helping to determine which other plants and sessile animals will grow beneath them. Beautiful animals such as sponges and sea-squirts can often be found growing beneath kelp forests, and studies have shown that the relationships among the organisms are very complex.

Humans, too, interact with kelp. Many kelps are cultivated or harvested for food, fertilizer, and as a natural source of products such as algin and potash. Because kelps grow close to the shore they are adversely affected by the human habit of disposing waste products in coastal waters. ●

Regular and evenly spaced, lobes line the frond edges of the giant kelp Macrocystis, *which occurs in many parts of the world.*

Kelp forests support very diverse communities of animals. Some species are restricted to them and others are more widespread, like this harbor seal, Phoca vitulina, *which is common also in other marine habitats across the northern Pacific and Atlantic oceans.*

the giant kelp, have floats or gas-filled chambers at the base of their fronds to maintain buoyancy, while others remain upright without floats. All plants are attached to the bottom by a root-like structure called a holdfast.

Kelps have a fascinating life cycle. They begin life as microscopic spores produced in special tissue on the mature adult plants, or sporophytes. When released, these spores develop into tiny male and female gametophytes. The male gametophytes fertilize the eggs produced by the females, which develop through embryonic and juvenile stages into the large marine plants. Adult kelps produce thousands of spores, but

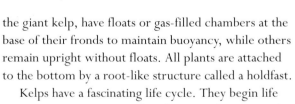

Pictor International/Austral

PART TWO

LIFE IN THE OCEANS

*The oceans of the world are teeming with life,
from microscopic plankton to the mighty blue whale.
Some species wander the oceans in search of food
or in response to migrational instincts,
whereas others are found only in specific habitats.
Yet despite many similarities, each ocean has
a distinct identity and range of species.*

4 OCEAN WANDERERS

MICHAEL M. BRYDEN, STEPHEN GARNETT, M.J. KINGSFORD, AND JOHN E. McCOSKER

From minuscule plankton to mighty whales, creatures of many different kinds migrate through or over the world's oceans. Some are long-distance voyagers, traveling vast distances in response to biological impulses. Others migrate through more limited areas, but with a tenacity that continues to amaze observers. Seabirds criss-cross the oceans, following a course that is as predictable as the season. Turtles and seals move between feeding and breeding grounds. Communities of plankton migrate vertically, moving up and down the water column in timeless natural rhythms.

THE OCEAN'S BUILDING BLOCKS

Plankton is a body of small plants and animals that drifts in the ocean. Plankters (members of the plankton) range in size from minuscule microbes to jellyfish with a gelatinous "bell" up to 1 meter (3 feet) wide and tentacles extending over 3 meters (10 feet). The Greek word *plankton* means "wandering or roaming", and although plankton is considered to drift with currents and tides, most plankters are anything but passive. In fact, many animal and some plant plankters undergo daily vertical migrations over tens or even hundreds of meters.

THE OCEANIC FOOD CHAIN

Phytoplankton ("plant plankton" in Greek) and zooplankton ("animal plankton") are important elements in all oceanic food chains. On land, through the process of photosynthesis, trees, grasses, shrubs, and flowers utilize

Opposite. Relatives of corals and sea-anemones, jellyfish are ocean wanderers like plankton, but move by rhythmic pulsations of the upper surface (bell), rather like an umbrella being open and shut. Sometimes they reach extraordinary densities, as in this congregation half-hiding a diver in the lagoons of Palau, Micronesia.

Claude Carre/Jacana/AUSCAPE

The great food chains in the sea all have their origin in plankton, tiny plant and animal organisms that drift near the surface in enormous numbers. This is Noctiluca miliaris, a minute bioluminescent protozoan.

sunlight, carbon dioxide, water, and nutrients to grow. The animals that eat them range from small insects to elephants. In the open ocean the plants are microscopic, but are often found in such high concentrations that the ocean can appear green. The main consumers of phytoplankton are microscopic animals.

The most productive regions of the oceans are where upwelling water from the ocean depths brings nutrients toward the surface. Phytoplankton thrives under these conditions, and grazing zooplankton multiplies with the abundance of food. Herring, sardines, and anchovy feed on the abundant plankton and are in turn preyed upon by large fish and birds. Some of the world's richest fisheries are found in areas of upwelling.

If upwelling does not occur, the effects on the animals that depend on plankton can be disastrous, and fisheries generally suffer as a consequence. The so-called El Nino phenomenon along the coast of Peru has been shown to result in a disruption to normal upwelling. In the longer term, scientists are concerned that changes in weather patterns caused by an increased concentration of greenhouse gases in the atmosphere will alter cycles of upwelling. The Earth's supply of atmospheric oxygen depends in part on phytoplankton, which, like other plants, releases oxygen as a byproduct of photosynthesis.

CYCLES AND PATTERNS

Cycles of plankton abundance vary in different parts of the world. In polar regions, plankton populations crash during the winter when there is constant darkness and an extended field of ice. In summer, with almost total light, plankton reaches peak abundance. In temperate regions there is usually a spring bloom of phytoplankton and a small bloom in autumn. Although concentrations of phytoplankton are low in tropical waters, high rates of reproduction by phytoplankton and of grazing by zooplankton make the turnover rate of plankton rapid.

Light does not penetrate sea water below depths of 200 meters (650 feet). Above this is the photic zone in which the plant life of the ocean photosynthesizes. Organisms that live on or near the bottom, below the photic zone, depend for survival on a "fall-out" of material from the upper layers. Plant cells, fecal pellets, microbes, live and dead plankton become bound together as mucus-like blobs called marine snow. In this form, food passes from the upper layers of the ocean to the bottom.

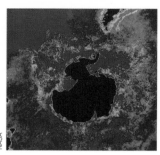

NASA

NASA

Seasonal phytoplankton blooms characterize polar waters; these satellite images show peak summer blooms in the Arctic (top) and Antarctic (bottom).

The oceanic food chain describes the systematic transfer of the sun's energy to phytoplankton and then as food to zooplankton and a host of intermediate animals, culminating in such large predators as killer whales and seals.

KINDS OF PLANKTON

The range of oceanic plankton is best described on the basis of size. The smallest kinds of plankton are bacteria with a minimum size of about 0.001 millimeter. Photosynthetic phytoplankters range in size from 0.002 millimeter to over 1 millimeter ($\frac{4}{100}$ inch). Only recently has it been recognized that plankters below 0.02 millimeter are responsible for a large proportion of oceanic primary production, especially in areas where nutrient levels are relatively low. Tiny bacteria gain some of their nutrition by clustering around phytoplankters that are actively photosynthesizing and utilizing organic material released by the phytoplankton.

Among the single-celled organisms that belong to the phylum Protista (in Greek, "first of all"), the diatoms and dinoflagellates are extremely important elements in the plankton. Diatoms are plant cells that range in size from a few thousandths of a millimeter to 1 millimeter ($\frac{4}{100}$ inch). Individual cells often form chains a few centimeters long, while dinoflagellates photosynthesize as plants and ingest solid food particles. Tiny flagellate cells consume bacteria and the smallest of phytoplankton. Other groups of protists include foraminifera, radiolarians, acantharians, and ciliates. Ciliates are an important source of food for recently hatched fish in the wild and are used in the aquaculture industry.

Zooplankters larger than 0.05 millimeter are grazers of phytoplankton, and some are predators. They are divided into two general categories: meroplankton,

which spend only part of their life as plankton, usually as larvae; and holoplankton, which spend their entire life as plankton. Many animals that are familiar to us in their adult form, such as crabs, barnacles, lobsters, and sea urchins, spend time as plankton, and these larval forms bear no resemblance to their adult appearance.

Important grazers and consumers of microbes include copepods, cladocera, larvaceans, and salps. Many holoplanktonic animals prey on other plankton: predators include jelly plankton and arrow-worms (chaetognaths). Predators can have a marked effect on some groups of plankton. For example, it has been argued that because jellyfish consume large numbers of larval fish, very high concentrations of jellyfish may reduce the size of fish populations. Some zooplankters are omnivorous: euphausids, or krill, for example, usually feed on phytoplankton, but will happily consume zooplankton if it is available as easily targeted packages.

MOVING THROUGH THE WATER

Despite their tiny size, most planktonic organisms migrate regularly through their environment. The best-known migrations of plankton—both phytoplankton and zooplankton—are vertical, where organisms move up and down in the water column. Movements of phytoplankton are generally achieved by controlling the amount of gas, oil, or salt within the organism. The production of gas or oil, or the removal of salt, will cause the organism to rise, while their release or absorption will cause the organism to sink. Some phytoplankters use mobile whip-like hairs called flagellae to help them move.

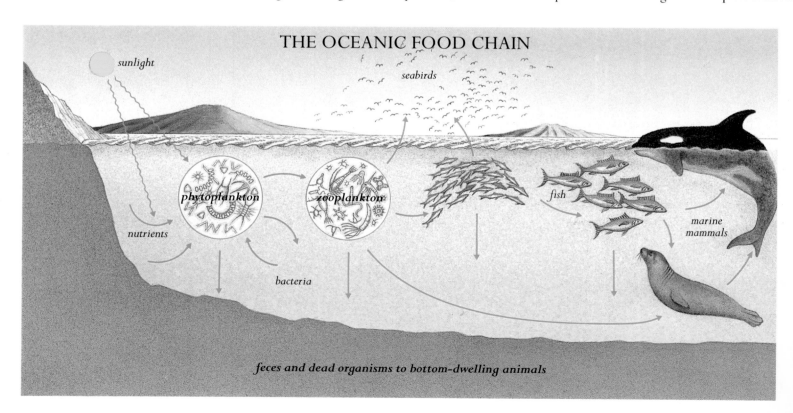

THE OCEANIC FOOD CHAIN

sunlight

seabirds

phytoplankton

zooplankton

fish

marine mammals

nutrients

bacteria

feces and dead organisms to bottom-dwelling animals

Three patterns of vertical migration have been recognized: nocturnal migrations, where plankton moves toward the surface at night and descends to deeper water during the day; reversed migrations, where plankton moves toward the surface during the day and to deeper water at night; and twilight migrations, in which organisms migrate at dawn and dusk. The most common pattern is nocturnal migration. The distance zooplankters migrate varies considerably and in the open ocean scientists have recognized a "ladder" of migrations. For example, some species may migrate only through the upper 100 meters (330 feet) of the water column, while others migrate over 1,000 meters (3,300 feet).

It is believed that many zooplankters move toward the surface at night to prey on food that is abundant in the photic zone; moving back to deeper water before dawn may reduce the chance of being eaten by predators. Moreover, some zooplankters also reproduce in surface waters. So the timing and duration of migrations may be regulated by the light–dark cycle and the presence or absence of prey, predators, or mates.

Horizontal movements of plankton are generally limited. However, by migrating vertically, plankters sometimes move considerable horizontal distances because currents often move in different directions at different depths.

ANCIENT PLANKTON

The skeletons of some plankters, such as the single-celled foraminifera and radiolarians, are well suited for preservation in sediments, and some of the best records of animals and plants that lived in ancient seas come from sediments and sedimentary rocks that contain plankton. Micropaleontologists examine sediments or thin sections of rocks to describe the microfossils they contain. Fossilized plankton has provided important information about plate tectonics. Sediments with relatively young fossils are found close to areas with active sea-floor spreading. The age of fossil plankton increases with distance from ridges where new sea floor is being produced through volcanic activity. Because we know the age of many fossil zooplankters, they are sometimes used to date rarer fossils found in the same rock layers.

The presence of some plankters in ancient seas is conspicuous to us even now. For example, white chalk cliffs in Europe, 65–100 million years old, are made up primarily of coccoliths—small planktonic organisms with calcareous skeletons. The breakdown products of ancient marine plants and animals have also contributed to reserves of petroleum trapped in geological formations.

M.J. KINGSFORD

Claude Carré/Jacana

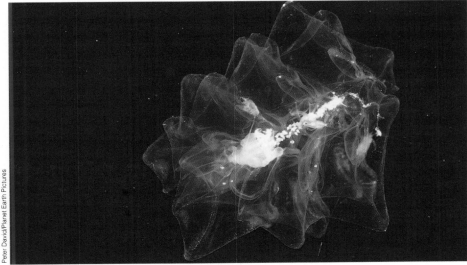

Peter David/Planet Earth Pictures

FISH: ADAPTATION AT ITS BEST

Like the Flying Dutchman, some fish are bound to spend their entire lives on the open sea. Others temporarily leave fresh water or the coastal margins in exchange for oceania, and do so for a variety of reasons: to avoid competition with their own kind by dispersal from their birthplace; to take advantage of a great feast; or to avoid the seasonal variations of a temperate climate in favor of a tropical one.

Evolutionary biologists who analyze life history strategies suggest that anything a species can do to improve its reproductive success is desirable, and that those individuals that live in a predictable and benign habitat will be more successful. So, if one's habitat is variable, often harsh and unpredictable, it should be traded for a better one—and that is just what many migratory and oceanic fish have done. Such fish are typically highly fecund species, early to mature, and often short-lived. Some migratory species have extended this to the extraordinary reproductive strategy of semelparity (suicidal reproduction). Coastal

Top. This strange planktonic animal is the phyllosoma stage of a Macrura reptantia, *and is scarcely recognizable as the larval form of a lobster. Phyllosoma larvae spend 11 to 13 months in the plankton and during this time undergo several molts.*

Bottom. Appearances can be deceiving: gelatinous zoo-plankters like this siphonophore are voracious predators of other zooplankton.

THE REMARKABLE TUNAS

JOHN E. McCOSKER

The fish that best represent adaptation to the oceanic environment are the tunas and their relatives. These 58 species are divided among three families within the suborder Scombroidei, which to a non-ichthyologist means the tunas, bonitos, seerfish, ceros, sierras, wahoos, mackerels, sailfish, marlins, and swordfish. They live in the surface waters of all tropical and temperate oceans, generally in the top 150 meters (500 feet) or so. They are well camouflaged with a dark blue-green dorsal surface, silvery sides, and a bright underbelly. Such a livery makes them difficult to see from above or below by presumptive predators or their own potential prey.

The tunas, billfish, and their relatives have attained the ultimate specializations for swimming in the open sea. Their bullet-shaped hydrodynamic bodies are so streamlined that they swim through the water without visible effort. In fact, tuna cannot stop swimming, for should they do so they would be unable to breathe. They have achieved this streamlined elegance via a Faustian bargain with their respiratory system: the muscle action required to pump water across their gills has been discarded in favor of swimming with their mouth open like a ram-jet ventilator. Their oxygen and energy demands are very high, needing a food intake of as much as one-quarter of their body weight each day. Their torpedo-like bodies are covered with small scales, and to reduce drag their median fins fit into grooves while swimming. This hydrodynamic efficiency allows the wahoo *Acanthocybium solandri* to reach 76 kilometers (47 miles) per hour and the sailfish *Istiophorus platypterus* to attain speeds of 109 kilometers (68 miles) per hour.

The muscle composition of the oceanic tunas is also designed for sustained swimming at high speed. The unusually high proportion of red to white muscle mass allows the northern bluefin tuna *Thunnus thynnus* to cross the Atlantic in 119 days—a distance of at least 7,770 kilometers (4,800 miles). In the Pacific, skipjack tuna *Katsuwonus pelamis* have been tagged and rediscovered nearly 10,000 kilometers (16,000 miles) away.

PACIFIC OCEAN

THE SKIPJACK TUNA

migration paths ——
spawning ground

Jeff Rotman

A diver brings up a bluefin tuna Thunnus thynnus *trapped in a net. Now seriously under threat, the bluefin tuna is one of the most important oceanic fishes harvested for commerce and sport. Bluefins may grow to more than 3 meters (10 feet) in length and up to 550 kilograms (around 1,200 pounds) in weight; such old giants tend to be solitary, but younger fish often occur in huge schools.*

Migration is simplified for most tunas through anatomical specializations which result in raised body temperatures. The heat generated by the red muscle mass is captured by special arterial and venous shunts which other fish lack, and thereby increases the temperature of the blood entering the gills and surrounding the viscera. This warm-bodiedness, rather than warm-bloodedness as higher vertebrates have achieved, allows for greater muscle efficiency and strength as well as the ability to inhabit a broad range of environmental temperatures. ●

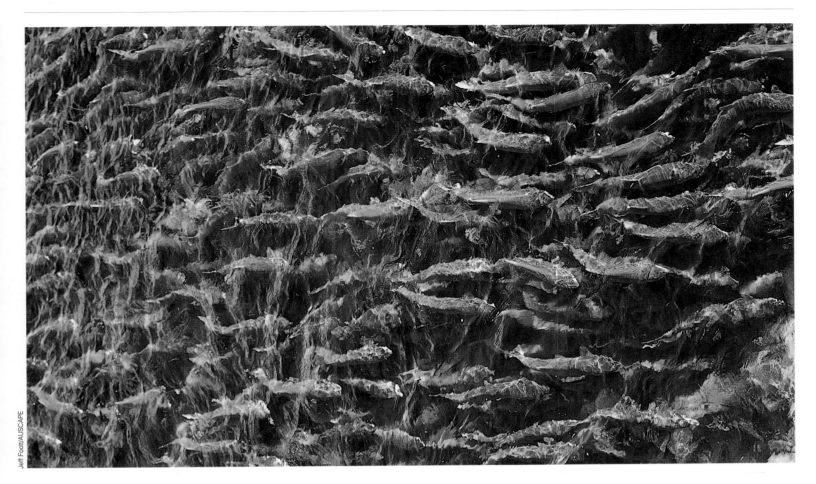

Jeff Foott/AUSCAPE

species that exist in more variable, unpredictable habitats often live longer, mature later, are less fecund, and reproduce repetitively (iteroparity).

The absence of strong geographical and topographical barriers to oceanic fish movement has limited their opportunities for speciation, and for that reason there are considerably fewer oceanic species than those that occupy the nearshore and fresh water. However those that have adapted to oceanic life have done so very well.

The phenomenon of migration from fresh water to the ocean, or vice versa, is called diadromy. Fish that spend most of their life in the sea and return to fresh water to breed—such as shads, sturgeons, and Pacific salmon—are said to be anadromous; diadromous fish that spend most of their lives in fresh water and migrate to the sea to breed, such as the freshwater eel, are called catadromous.

The best-known example of anadromy is that of Pacific salmon of the genus *Oncorhynchus*. These fish leave the coastal waters of the north Pacific and cross the ocean, often as far as Japan, returning several years later to their precise home stream to spawn and die. Biologists have discovered that they probably navigate by means of an internal compass, aided by geomagnetic clues. Once within their home river system, they complete their journey by following chemical cues and odors which are unique to their parental spawning site and are recalled from imprinting at the time they were fingerlings.

Anadromy is more common in temperate waters, and is presumably related to the ratio of available food.

Catadromy is a rarer and tropical event. The Atlantic freshwater eels, *Anguilla anguilla* and *A. rostrata*, undergo an incredible and, as yet, poorly understood migration. The adults leave the Mediterranean and the Americas and congregate in the deep water of the Sargasso Sea. There they breed and die, spawning the eel larvae which will ultimately return to the American and European rivers. Diadromous fish, particularly the anadromous species, are not numerous in species, but are so in biomass, and for that reason their effect upon the food chain is significant.

These strategies are a marvel of the process of natural selection, but are unprepared for the development of high-seas fisheries. The absence of geopolitical boundaries and the increasing demand for protein increases the difficulty of stock assessment and management. International cooperation will be required if oceanic and migratory species are to be conserved.

Sockeye salmon Oncorhynchus nerka *migrate in vast numbers from the sea to their spawning grounds in the gravelly upper shallows of rivers and streams in western North America. This is their final journey—all die within a few days of laying and fertilizing their eggs.*

During spawning the male sockeye salmon (right) takes on a brilliant red color and develops a long, hooked jaw and sharp teeth, which it uses to battle for females. The development in females (left) is less dramatic.

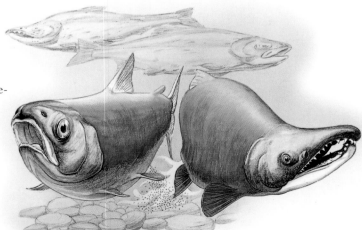

The bewildering intricacy and diversity of the world of zooplankton is evident only under the microscope.

Opposite. *A gentle giant, the world's largest fishlike creature bypasses the normal oceanic food chains by feeding directly on plankton. Solitary, uncommon, and inoffensive to humans, the whale shark may grow to 12 meters (40 feet) in length— though some have been reported at sea, plainly larger but unmeasured, up to 18 meters (59 feet) long.*

OCEANIC SHARKS

It is commonly presumed that the open ocean is rife with sharks. Few would disbelieve Herman Melville when, in *Moby Dick*, he described the abundance of sharks around the *Pequod* at anchor: "Any man accustomed to such sights, to have looked over her side that night, would have almost thought the whole round sea was one huge cheese, and those sharks the maggots in it."

But now, limited food resources, a reproductive behavior which limits their capacity to respond to population pressures, and the onslaught of high-seas shark fishing have reduced their numbers.

There are about 350 living species of sharks, and of them, only five or six species are truly oceanic. Most are coastal species and a few inhabit the deep-sea bed. Even if one includes the 425 species of skates, rays, and sawfish (considered by modern ichthyologists to be merely flat sharks), only one or two additional oceanic wanderers may be added.

Truly oceanic sharks range in size from the poorly known crocodile shark *Pseudocarcharias kamoharai*, which grows to a mere 110 centimeters (43 inches), to the whale shark *Rhincodon typus*, by far the world's largest fishlike creature, which possibly reaches 18 meters (59 feet) in length. The only other sharks that primarily occupy the high seas are the blue shark *Prionace glauca*, the shortfin mako *Isurus oxyrinchus*, the oceanic whitetip shark *Carcharhinus longimanus*, and the silvertip shark *Carcharhinus albimarginatus*.

Like the great whales, the whale shark is so large that it is incapable of capturing almost any sizeable prey. So it too has turned to feeding on abundant and microscopic phytoplankton and zooplankton. In so doing, each one probably consumes a considerable biomass. The whale shark is not common, and is rarely seen in a group. It is tolerant of humans, occasionally curious, and harmless.

The mako shark, however, is a fearsome creature, and were it not for its offshore habitat, it would often be in the headlines for consuming humans. It is fast and powerful, has large, sharp, and bladelike teeth, and grows to nearly 4 meters (13 feet). It is a common offshore species in tropical and warm temperate waters, and can appear almost anywhere if the water is warmer than 16°C (61°F).

Other dangerous oceanic sharks include those of the genus *Carcharhinus*. The silvertip and the oceanic whitetip can also be found in most tropical seas, but fortunately live offshore. Due to the scarcity of resources on the high seas, they consume bony fish and squids, birds and turtles, and carrion of almost any size or kind. With their long, broad, paddlelike pectoral fins, the whitetips are most impressive. They swim

Top. *A researcher investigates the behavior of a blue shark from the safety of a metal cage. One of the most common and widespread of sharks, the blue shark Prionace glauca inhabits offshore and deep-ocean waters in temperate and tropical regions the world over.*

Bottom. *A member of the whalers, or requiem shark family, Carcharhinidae, the silvertip shark Carcharhinus albimarginatus usually frequents the seaward ridge of coral reefs, especially at depths of more than 25 meters (82 feet).*

languidly at or near the surface and have been observed to enter calmly a school of small, frenetically feeding tuna, open their capacious maw, and wait for a tuna to swim recklessly in. The strength of their jaws and the sharpness of their teeth were well known to whalers who often saw their floating catch dismembered by whitetips or silvertips.

And finally, the blue shark, probably the most abundant and adaptable of living sharks. It was originally named *Squalus glaucus* by Linnaeus in 1758; its type locality "Habitat in Oceano Europaeo". This graceful, dark blue shark reaches about 4 meters (13 feet) in length, and travels in all temperate and tropical seas.

It is often found in large groups and occasionally near the shore. Tagging studies of blue sharks demonstrate that they ride the clockwise currents in the north Atlantic, crossing from America to Europe with the Gulf Stream and returning to the Caribbean by the westward-flowing North Equatorial Current. In the Pacific they seem to move south in the northern winter and northward in the summer. They too will eat almost any protein they confront, but are limited by the smaller size of their mouth and jaws.

Blue sharks are quite fecund, as elasmobranchs go, and upon reaching five years of age females can give birth to as many as 135 young. But their abundance is declining worldwide as a result of the heavy fishing pressure wrought by humans. Tens of thousands die annually in oceanic barrier nets as an incidental bycatch of more valuable fish. An increased need for protein, as well as the epicurean desire for sharkfin soup, have now directed fisheries in search of sharks, rather than killing and discarding them as pests. Population data for oceanic sharks are difficult to obtain, but necessary if fisheries are to continue without causing major and irreversible depletion of the stocks. International cooperation is necessary but will be difficult to achieve for creatures of the sea that so repulse and frighten us.

JOHN E. McCOSKER

WHALES AND DOLPHINS

The distribution and migration of animals is governed largely by their food supply. In the oceans, food for whales and dolphins is quite patchy; in some areas fish, cephalopods, crustaceans, and other organisms abound, while large tracts of ocean are virtually barren. For the most part, tropical seas are relatively poor in nutrients. Polar seas contain up to one hundred times the concentration of organisms found in most tropical oceans.

A few areas in the tropics are, however, very fertile and support teeming masses of plankton, fish, and squids—for example, the sea off the Galapagos Islands, the Caribbean Sea, and the Arabian Sea. In some tropical regions, upwelling brings nutrients to the surface to support a complex network of organisms. Very large schools of dolphins, sometimes in the thousands, congregate here, some remaining for most if not all the year, together with such predominantly tropical whales as Bryde's whales *Balaenoptera edeni* and female sperm whales *Physeter catodon*.

Many species of whales and dolphins are local, living over or near the continental shelf of landmasses, or very close inshore, and in sheltered bays and inlets. They live singly or in small groups, and make up a relatively small total biomass which their prey (mainly fish, cephalopods, crustaceans, and other small invertebrate species) can sustain.

Norbert Wu/Planet Earth Pictures

Jack Jackson/Planet Earth Pictures

Before this century, when pelagic whaling using factory ships was introduced, the great whales, including blue, fin, sei, and humpback whales, were abundant and made up a huge biomass. These are tropical whales that mate and give birth in the tropics, usually in areas in which there is little or no food. They make regular annual migrations from tropical breeding areas to feeding areas in high latitudes. The evolution of this migratory pattern is uncertain, but it must be related to their having ranged from their breeding areas and found sources of food in colder waters, probably beginning in the Tertiary period about 30 million years ago. Climatic and other changes possibly led to an increased geographical separation between breeding and foraging areas.

The cold-water oceans that are so rich in small organisms can sustain a massive biomass of predators, which include the great whales, seals, seabirds, and large fish. The abundance of food in these polar seas enabled the whales to increase in individual size and total numbers. Many of them have become highly specialized, particularly with regard to their feeding apparatus. The great whales lost their teeth and developed baleen plates, the horny downgrowths from the palate, frayed at their tips, that together form a "curtain" to filter out the small organisms that make up the whales' diet.

A pod of humpback whales Megaptera novaeangliae *feeding in Alaska. Humpbacks are strongly social animals with a close-knit group structure. They feed largely on schooling fish, which they often circle with characteristic breaching behavior to concentrate the panicked fish more densely.*

Bill Wood/Australian Picture Library

Small, playful cousins of the great whales, porpoises and dolphins inhabit mainly coastal waters, though there are several truly oceanic, deepwater species. This is the bottle-nosed dolphin Tursiops truncatus, which has an almost worldwide distribution in temperate and tropical seas, and is the species most widely kept in zoos and oceanaria.

The annual migrations of the great whales involve some risk. In returning each year from the polar seas to breed in their tropical "home", they leave a feast to migrate through vast tracts of ocean in which there is very little food, to reach breeding areas where there is virtually none. Adult females are particularly vulnerable, because while fasting they support a rapidly growing fetus on the way to the tropics, and must produce milk for a sucking calf as they venture back to the polar seas. They lose a vast amount of weight, and the stress is undoubtedly too much for some.

Two questions about these phenomenal migrations beg an explanation: why do these whales, that can withstand cold so well, return to tropical waters to breed? And why is food so abundant in polar seas?

The answer to the first question is probably that in cold waters at high latitudes the food supply is reduced in the winter months when the sea freezes over, and the small organisms that make up the zooplankton descend to greater depths and virtually "hibernate". In order to maintain their body temperature, whales probably have to move about to some extent, and energy loss is considerably less in the tropics than in the freezing polar seas. Moreover, because the newborn whale calves have a relatively large skin surface area and virtually no blubber to insulate them against the cold, they may die from body

heat loss in freezing water. However, some whale species, including white whales and narwhals, have managed to conquer this problem somehow, and do remain in the cold Arctic sea year-round.

The answer to the second question is more straight-forward. The enormous living resources of the polar seas result from several oceanographic factors. There is a drift of nutrients along the sea floor toward the poles, where the nutrients rise to the surface and move toward warmer water. So the surface layers of ocean near the poles are extremely rich in the nutrients required by phytoplankton. With the help of sunlight, these synthesize organic matter from carbonic acid and water. Carbonic acid is more soluble and oxygen is more plentiful in cold than in warm water. Conversely, destructive bacteria thrive in warmer temperatures. Consequently, the ingredients are present in polar waters to permit mass blooming of phytoplankton in the summer months, when sunlight is available for photosynthesis almost 24 hours a day. The phytoplankton is the foundation of an enormous crop of zooplankton on which the whales and other large predators feed. But in winter, when there is little sunlight, the phytoplankton stops growing, the surface water freezes to form sea ice, the zooplankton descends in the water column and virtually stops developing. Thus it is unavailable to predator species.

D. Parer & E. Parer-Cook/AUSCAPE

THE ELUSIVE TURTLE

Although sea turtles are the most spectacular migrants among the reptiles, information about the distribution and migration of various species is not obtained easily, and the only reasonably detailed knowledge refers to nesting populations of the green turtle *Chelonia mydas* that breed along the coasts of Costa Rica and Ascension Island.

Eggs are buried as they are laid, above the high-tide line on sandy shores of tropical and subtropical mainland and island coasts. Some six to eight weeks later the young hatch and immediately migrate across the sand to the sea, into which they effectively disappear. Nothing is known about where they go or what they feed on during their first year of life. It is assumed that, as they leave the beaches where they were born, they drift with the ocean current for a period. This period may be short, perhaps only a few weeks, during which the young turtles arrive in nursery areas where they may live among the floating rafts of *Sargassum* weed. A more likely explanation, however, is that they live a pelagic existence for a year, feeding on zoo-plankton as they go. Even when they adopt an adult-type herbivorous diet,

young green turtles do not remain long in one place, and their migrations could well be exploratory while they learn about their environment and establish a familiar area.

Adult green turtles feed predominantly off continental coasts on turtle grass, an underwater flowering plant. At maturity they select an appropriate beach for nesting, and the second phase of the lifetime migration circuit begins. Grazing and breeding areas may be 2,000 kilometers (1,250 miles) apart. Every two or three years females leave the feeding areas, migrate to the breeding areas, lay from three to seven batches of 100 or so eggs at 12-day intervals, and then return, migrating with the surface current, to the grazing areas. The migratory pattern of males is less certain, but undoubtedly occurs, because males are also seen around the breeding areas. Copulation takes place offshore, when the eggs that will be laid two or three years later are fertilized.

Adult green turtles Chelonia mydas *live solitary lives, feeding on seaweeds or seagrasses. Every two to eight years they migrate to breeding areas, often several thousand kilometers from their home feeding grounds.*

Female green turtles come ashore about two weeks after mating in search of suitable spots to excavate nest cavities. Each female returns to lay several times during a season, at about two-weekly intervals. Several thousand turtles may be on some beaches at once, and many millions of eggs are laid over the entire nesting season.

Collectively known as pinnipeds, the seals, sea lions, fur seals, and walruses form a group inhabiting mainly coastal waters from the equator almost to the poles. Although all pinnipeds feed in the sea, most also spend a good deal of time loafing ashore. The California sea lion Zalophus californianus (below right) is also common in the Sea of Cortez and in the Galapagos. The northern fur seal Callorhinus ursinus (below) inhabits the fringes of the Bering Sea, whereas the southern elephant seal Mirounga leonina (far right) occurs in subantarctic regions.

Solitary Travelers

All seals, sea lions, and walruses, known collectively as pinnipeds, are amphibious, alternating between periods in the water and periods on land, on mainland coasts, islands, sandbanks, ice fields, or ice floes.

For each species, or among breeding groups within species, births are concentrated into a relatively short time of the year, a period that is followed fairly quickly by the main mating season. An exception is the Australian sea lion *Neophoca cinerea*, which has a protracted gestation and breeding season, and is believed to have a breeding cycle that is 18 months long.

Dispersal in pinnipeds is associated with feeding; convergence is associated with birthing, copulation, molting, and resting. Pinniped species such as the northern fur seal *Callorhinus ursinus* and the harp seal *Phoca groenlandica*, that are distributed at well-marked places during the breeding season, disperse after breeding as individuals migrate different distances in different directions. Species like the southern sea lion *Otaria byronia*, which breeds along continental coasts, and the crabeater seal *Lobodon carcinophagus* and leopard seal *Hydrurga leptonyx*, which breed among or at the edge of the circumpolar band of Antarctic pack ice, disperse when individuals travel different distances rather than in different directions.

John Shaw/AUSCAPE

David Doubilet

Some species, such as the leopard seal and the Ross seal *Ommatophoca rossii*, are more or less solitary throughout their lives. Other species, such as the northern fur seal and the elephant seals *Mirounga* are intensely gregarious during the breeding season, but they are quite solitary while at sea, though a certain amount of gathering does seem to occur when food is abundant.

There is a suggestion that species which feed predominantly on fish and cephalopods may disperse more than other species, and they may also be more inclined to have individual home ranges. On the other hand, species which feed on invertebrates such as krill or bottom-dwelling mollusks and crustaceans may show less dispersal. Those species that breed on oceanic islands seem to show greater convergence during the migration to the breeding grounds whereas coastal breeders and ice breeders seem to show progressively less convergence during this phase of the migration.

Until recently, most of what was known about the migration of seals had been learned from resightings of marked animals, and there were, therefore, large gaps in our knowledge. In the past few years the fitting of time-depth recorders and, more recently, satellite recorders, to individual animals has given us much greater insight into the distribution, range, and migration of some seal species.

MICHAEL M. BRYDEN

Jean-Paul Ferrero/AUSCAPE

OCEANIC BIRDS

The first birds evolved on land about 160 million years ago. Even as these feathered reptiles were learning to flap from tree to tree, toothed versions of today's loons were learning to swim. Birds have been associated with the sea ever since. Their use of the marine environment is constrained by two requirements: air to breathe, and land on which they can nest.

NESTING AND BREEDING

Characteristically, seabirds nest on islands where they are safe from land-based predators. At the height of breeding there can be few more spectacular sights in the natural world than an island of nesting birds, an exhilarating mix of noise, smell, and sheer fecundity. Some birds nest in burrows, some in dense colonies on the surface, and others on cliff ledges, laying eggs much narrower at one end than the other to prevent them from rolling off.

Several hundred species of birds exploit the ocean's surface in a variety of ways. Some are illustrated here.

These aggregations of birds have a profound effect on the islands on which they breed, concentrating nutrients from the surrounding seas into mountains of excreta or guano, which is mined as a source of fertilizer. Particularly famous are the islands off Peru where mountains of guano 90 meters (300 feet) thick had accumulated over thousands of years before they were carried away to Europe and America during the nineteenth century. Today, in both Peru and southern Africa, the removal of guano from such places is regulated.

Because such productive islands are home to tens of millions of breeding birds, a prolific source of food is needed nearby. This is especially true for colonies of penguins—birds that have become so well adapted to the sea that they have paddles for wings. Because they cannot travel as far as flying birds, penguins need to catch nearly all the food for their young within 20 kilometers (12 miles) of their breeding colonies. Their only compensation is that they can dive deeper than other birds. The largest species, the emperor penguin

HOW SEABIRDS CATCH THEIR PREY

frigatebird

skua

Like frigatebirds but normally found in colder waters, skuas and jaegers harass other seabirds until they part with their catch.

gannet

phalarope

cape pigeon

storm petrel

Prions and some storm petrels feed largely on plankton, close to the surface.

albatross

tern

Frigatebirds divide their hunting between snatching flying fish and pirating the catches of other birds.

diving petrel

cormorant

penguin

auk

Penguins, auks, diving petrels, and some shearwaters actively pursue fish underwater, propelling themselves with their wings.

Cormorants hunt underwater like penguins, but use their feet for propulsion.

Aptenodytes forsteri, can feed at depths of more than 250 meters (800 feet). All other seabirds are confined to the top 20 meters (65 feet) of ocean.

FINDING THEIR WAY

Flying seabirds are more mobile than penguins. Some leave their nests for days or weeks at a time while they tour the ocean in search of food. In 1989 a wandering albatross *Diomedea exulans* was tracked by satellite over a distance of 12,000 kilometers (7,500 miles), flying from the Crozet Islands in the southern Indian Ocean to the Antarctic continent and back before relieving his mate at the nest. Other species may well travel as far.

It seems remarkable to us that seabirds can find their often tiny nesting islands in such a vast expanse of ocean. Birds, however, can probably pinpoint their position from the pattern of waves, winds, and currents just as we might read a map. Seabirds also have a strong sense of smell, which they use to find both food and their nesting sites when, as many do, they return after dark.

D. Parer & E. Parer-Cook/AUSCAPE

A group of emperor penguins Aptenodytes forsteri *brood their chicks. This extraordinary bird builds no nest. It breeds during the Antarctic winter, the male holding the single egg against the warmth of his abdomen with his feet until it hatches, whereupon the female returns from the sea to give the chick its first large feed.*

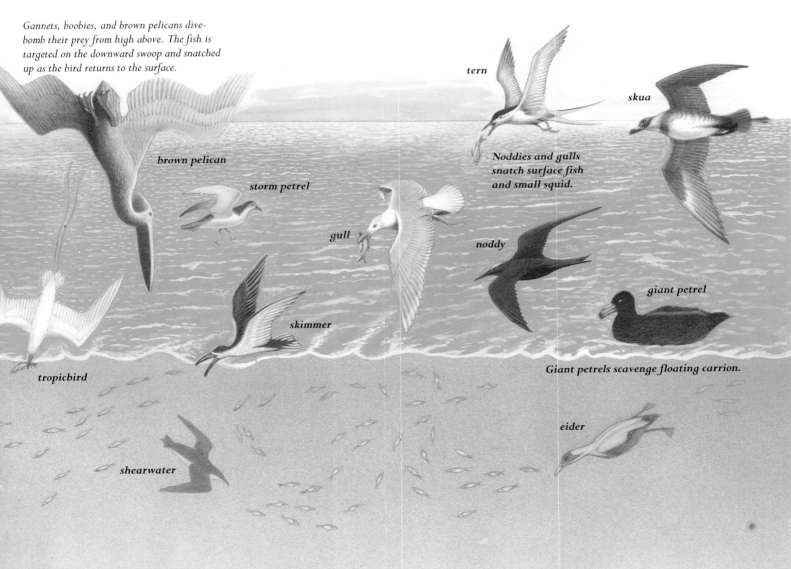

Gannets, boobies, and brown pelicans dive-bomb their prey from high above. The fish is targeted on the downward swoop and snatched up as the bird returns to the surface.

tern

skua

brown pelican

Noddies and gulls snatch surface fish and small squid.

storm petrel

gull

noddy

giant petrel

skimmer

tropicbird

Giant petrels scavenge floating carrion.

eider

shearwater

Opposite. *A pair of wandering albatrosses* Diomedea exulans *in courtship display at the nest site. These huge birds spend much of their lives wandering the open ocean, coming together only to breed. They mate for life, reinforcing their pair bond by elaborate and highly ritualized displays at the nest.*

The Arctic tern Sterna paradisaea *has the longest migration route of any bird: it breeds in Arctic regions, but crosses the equator to spend the non-breeding season in the pack ice fringing the Antarctic continent.*

This homing ability is all the more extraordinary considering the distances traveled by seabirds on their migration paths. The Arctic tern *Sterna paradisaea*, for instance, breeds north of the Arctic Circle between May and August, then travels the length of the globe to spend from November to March in the pack ice of Antarctica. As a reward it lives in almost constant daylight. Some other seabirds, such as shearwaters and storm petrels, migrate similar distances but breed in the Southern Hemisphere then pour north in tens of millions as the days shorten. Others move laterally, some albatrosses and petrels in the Southern Hemisphere simply following the wind. Unimpeded by continents, they probably circumnavigate Antarctica several times between breeding attempts.

DIVERSITY AND ADAPTATION

All seabirds are carnivores, the majority taking fish, squid, or krill. To find their prey, seabirds must be particularly sensitive to water temperatures and currents. Some species favor warm water, others cool. Those of the southern seas sometimes follow the Humboldt or Benguela currents, formed of cold, upwelled water rich in nutrients, which run northward along the western coasts of South America and southern Africa respectively. Often, productivity is greatest where waters of different temperature meet, particularly where cold, nutrient-rich water wells to the surface at the edges of continental shelves or in turbulence.

Especially rich are the cold waters of higher latitudes, and these support the greatest diversity of seabirds. Of the four major groups of birds adapted

THE ARCTIC TERN: A LONG DISTANCE WANDERER

ATLANTIC OCEAN

migration paths
wintering areas

to life at sea, two are concentrated in the Southern Ocean, one around the Arctic, and only one in the much larger area of tropical sea. In the south live all but one of the 17 penguin species and most of the so-called tubenoses, the petrels and albatrosses. In the Northern Hemisphere there is a great diversity of larids, the order of birds containing gulls, terns, puffins, murres, auklets, and guillemots. The group with greatest variety in the tropics is the pelecani-forms, which includes the pelicans, boobies, tropicbirds, frigatebirds, and shags. These groups have adapted to their marine environment in different ways, though they have in common webbed feet for swimming and salt glands for removing excess salt from their blood. The most aquatic birds are the flightless penguins, though some of the auklets and murres in the north and diving petrels in the south flap their tiny wings under the water just as they do above it. By contrast, the petrels and albatrosses have the most efficient gliding flight of any bird. Their long, narrow wings, as much as 3 meters (10 feet) from tip to tip in the wandering albatross, enable them to ride ocean winds effortlessly.

The boobies and tropicbirds are streamlined for plunging into the ocean, sometimes closing their wings and falling vertically through the air for 30 meters (100 feet) or more to capture fish. They, like most other seabirds, have a waterproof plumage, but their close relatives, the frigatebirds, have apparently sacrificed this capacity to save weight and energy. As a result they can stay airborne for weeks at a time, snatching food from the ocean's surface or stealing it from other seabirds.

THREATS AND PROBLEMS

Despite the expanse of their oceanic environment, many seabirds are threatened. Coming to land only on remote islands, most are absurdly tame, and several species have been exterminated by human exploitation for food or feathers. Introduced predators, such as rats and cats, have caused the extinction of many others. Now there are also threats at sea.

Thousands of birds die annually in gill and drift nets on the open ocean, and albatross populations have been declining steadily for the last 20 years as a result of longline fishing. Others are poisoned by pollution. Oil removes the waterproofing of the feathers, so birds freeze to death. Even Antarctic penguins show traces of pesticides, and nearly all seabirds contain plastic in one form or another that has been pecked from the ocean surface. Birds are such an important part of the marine ecosystem that it would be tragic if they were decimated by what is essentially human carelessness. ■

STEPHEN GARNETT

5 A DIVERSE HABITAT

RICHARD HARBISON, KNOWLES KERRY, J.R. SIMONS, AND DIANA WALKER

*D*espite many similarities and the mixing of waters along their boundaries, each ocean has a
distinct identity. The Pacific, Atlantic, Indian, and Southern oceans are typified by their physical
nature—temperature and salinity, for example—and biological characteristics. Similarities between the
oceans, such as the transfer of energy through food webs which encompass microscopic plankton through
to large marine mammals, ensure that life is maintained. Differences—often resulting from the varying
geological history of each ocean—make life in the oceans particularly fascinating.

*Squid-like in appearance but with
a snail-like shell, the pearly
nautilus* Nautilus pompilius
*inhabits deep waters, migrating to
the surface to feed at night. The
spiral shell is filled with gas and
divided into chambers linked by a
fleshy tube, which allows the
nautilus to adjust its buoyancy.*

Mike Tinsley/AUSCAPE

THE PACIFIC OCEAN

The diversity and abundance of life in the ocean
depend ultimately on the basic fertilizing nutrients
available in its waters. Productive surface waters
occur in the northern and southern extremities of
the Pacific, where the near-freezing water welling
up from the ocean floor carries prodigious quantities
of fertilizing minerals. In spring and summer the
waters bloom with millions of tonnes of microscopic
plankton upon which even large creatures such as
baleen whales or basking sharks can directly feed.
Apart from the microscopic organisms which photosynthesize carbohydrates, plankton also consists of
the eggs, larvae, or young of a wide variety of
marine creatures.

Oceanic currents sweep the rich cold waters from
the extremities of the Pacific toward the Americas
and the equatorial regions. Thus, in terms of fishery
economics, the northern, southern, and eastern
regions of the Pacific are the most productive. This
is not to deny, however, the economic importance
of the western regions or the biological interest of
the myriad forms of marine life which inhabit the
seas that surround the islands, coral reefs, atolls,
and cays that are so characteristic of the western
Pacific Ocean.

The waters of the northern and southern halves
of the Pacific tend to remain separate and many of
the creatures inhabit only one of the ocean's two
main systems. Some salmon, for instance, including
the prized *Oncorhynchus*, occur naturally only in the
north Pacific. On the other hand, widely roaming
oceanic wanderers such as the giant whale shark, the
Pacific marlin, and some dolphins and whales freely
cross from one hemisphere to the other.

CORALS AND REEFS

The Pacific immediately conjures up visions of coral
islands and reefs. These coral ramparts, built of
immense quantities of limestone, are produced by
innumerable diminutive relatives of the jellyfish and

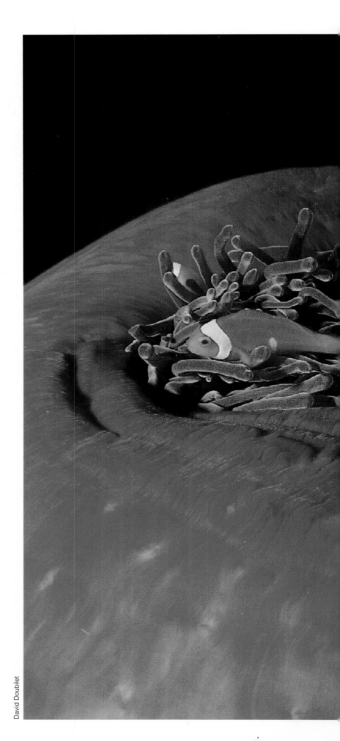

David Doubilet

sea-anemones, the so-called coral polyps. Each coral organism has the capacity to secrete some form of protective tube about its body. As it grows, it buds off younger individuals which remain attached to it, thus rapidly forming a colony of individuals joined with a common flesh. In some species—the soft corals—the protective coating is a tough, fleshy material. As a result, the whole colony may be quite flexible and, at first sight, could be mistaken for a variety of kelp.

In general, reef-building corals occur in the warmer waters of the ocean with the active part of the reef never extending more than about

37 meters (120 feet) below the surface of the ocean. Solitary corals, which resemble large anemones but with a limy exoskeleton, are generally found in deeper, cooler Pacific waters.

Corals and all the members of the zoological group to which they belong—including jellyfish, box-jellyfish, Portuguese men-o'-war, and sea-anemones—have tentacles that are equipped with microscopic stinging cells capable of ejecting a poisonous barb into the flesh of any potential prey that touches them. Any surfer who has been stung by the long trailing tentacles of the Portuguese man-o'-war *Physalia* (also known as the bluebottle) can attest to a

Clownfish of the genus Amphiprion *live in a close relationship with sea-anemones, sheltering in the protection of the anemones' stinging tentacles. Sea-anemone tentacles secrete a chemical that inhibits the stinging response so that they don't sting each other. Skin secretions on the clownfish closely mimic this chemical, so the clownfish shares the anemone's own defense against itself.*

David Doubilet

A tiger moray eel in Japanese waters. Most of the 80-odd species of moray inhabit tropical seas, especially coral reefs, where they hide in crevices during the day and emerge to feed at night.

Opposite, top. Schools of lutjanids, like these at Palau, Micronesia, inhabit coral reefs throughout tropical waters. Carnivorous and inquisitive, they are commonly known as "snappers".

Opposite, bottom. Widespread in Australasia, the blue-ringed octopus Hapalochlaena *differs from most octopuses, generally masters of protective coloration, in that its color pattern of dull ocher and electric blue is more or less constant, perhaps as a warning of its poisonous bite.*

painful experience. In the more tropical waters of the Pacific, various box-jellyfish (or sea-wasps) *Chironex* have venomous stings which have killed humans within minutes.

A coral reef harbors an amazing variety of fish, crustaceans, sea-urchins, starfish, brittlestars, mollusks, and worms. One of the most famous of the worms is the palolo worm *Leodice* of the waters around the coral-fringed islands of the central Pacific. As with many of its relatives, the eggs and sperm develop in the rear half of the creature's body. At the appropriate time, the rear section detaches from the parent body and swims up to the surface of the water. The remarkable thing about the palolo's swarming, as it is known, is the precision of its timing. It occurs on the day of the last quarter of the October–November moon. The swarming is so dense that it is easily harvested to provide the islanders yearly with a delectable feast.

MOLLUSKS AND CRUSTACEANS

Innumerable species of mollusks inhabit the Pacific, many of them unique to specific regions. The rock oyster *Crassostrea* of the eastern Australian coast is famed for its superior taste and for the fact that, unlike many other oysters, its offspring hatch and develop away from the parent.

Three other mollusks unique to the ocean are all members of the squid/octopus group. One, the blue-ringed octopus *Hapalochlaena* of the south-western region, although comparatively small, has a highly venomous bite. The second, the Humboldt squid in the waters off South America, is the largest of the Pacific forms, weighing in at some 50 kilo-grams (110 pounds).

Perhaps the most interesting of the three is the pearly nautilus *Nautilus pompilius*, which inhabits the waters of the coral reefs and islands of the south Pacific. While resembling the squid or octopus in having a head surrounded by tentacles, it also possesses a stout, spirally coiled, lustrous shell into which the animal can completely retract. With this large chambered shell, the creature resembles the long-extinct ammonite.

Crustaceans, particularly prawns and lobsters, are important human food resources. Pacific lobsters lack the claws of their northern European counterparts but have a pair of stout, spiny antennae which, when lashed about, are effective at deterring an enemy. While not unique to the Pacific, various species of prawns are certainly its most characteristic and widespread crustaceans. These too differ from the prawns and shrimps of European waters in several ways. Apart from anatomical points and the fact that they grow much larger, the principal difference is that, when they spawn in the deep sea, their fertilized eggs are set free and not stuck to the body of the female, as is the case with other types of prawns found in European waters and, indeed, crustaceans in general.

SHARKS AND RAYS

Sharks evoke less than romantic images of the Pacific. While some deserve their bad reputation, most seem to be blameless. The Pacific harbors a number of archaic creatures such as the Port Jackson shark *Heterodontus* and the frill-gilled shark *Chlamydoselachus*. The several species of the former live inshore of the coasts of eastern Australia and neighboring islands while the latter inhabits the waters of the north-western Pacific. *Heterodontus* is a harmless, mollusk-eating creature with powerfully muscled jaws for crushing the shells of its principal food. With its elongated sinuous body, *Chlamydoselachus* resembles an eel more than a shark. Fossil records, which for sharks extend 100 times as far back as those for humans, indicate that both sharks have a long geologic history, being almost unchanged in form since Jurassic times some 150 million years ago.

Closely related to sharks are the rays. The largest of these unique to the Pacific is Captain Cook's stingaree *Bathytoshia*, which reaches lengths of over 4 meters (13 feet) and can weigh more than 200 kilograms (440 pounds). Larger still is the more widely distributed manta ray, often referred to as the devil fish despite the fact that it is a non-aggressive plankton feeder.

MARINE MAMMALS

Some 20 species of whales, dolphins, and porpoises are exclusive to the Pacific Ocean. They do, however, share the ocean with some of the other more widely distributed species. Of the 20 species endemic to the Pacific, there is one baleen whale, the Pacific right whale *Eubalaena*. The remaining

Pictor International/Austral

D. Parer & E. Parer-Cook/AUSCAPE

GUARDIAN OF THE KELP BED

Australia. Until the nineteenth century, it had a close relative in the north Pacific, Steller's sea-cow, which became extinct from over-hunting.

Another Pacific mammal is the sea-otter, which occurs in the colder waters of the American shores of the north Pacific. The largest of the otters, over 1 meter (3 feet) long, it rarely leaves the water and, to rest or sleep, beds itself in surface kelp or seaweed.

Another species, not quite as fully marine, but having the same general behavior, lives far to the south in Tierra del Fuego. Sea-otters have the remarkable habit of fetching a flattened stone from the depths and, while floating on their back at the surface, placing the stone on their chest. They then proceed to open food such as sea-urchins, mollusks, clams, and crabs by cracking them against the stone.

J.R. SIMONS

Sea-otters Enhydra lutris *inhabit coastal waters of western North America from California to the Aleutian Islands, especially in kelp beds along rocky shores. They dive for sea urchins and similar bottom-dwelling animals, bringing them back to the surface to eat. They often use a flat stone, laid on the chest, as a sort of anvil against which to break open their prey. Sea-otters are strongly aquatic, coming ashore only during exceptionally heavy storms, or to give birth to their pups.*

19 are all toothed creatures which actively pursue their food. It is a mixed group consisting of five species of beaked whales, 11 of dolphins, and two of porpoises. The nineteenth is the Pacific pilot whale *Globicephala*. This whale was the first of the large whales to be kept and trained in captivity.

In the Southern Hemisphere, seals are generally regarded as creatures of the polar region. Nevertheless, the leopard seal *Hydrurga* occasionally appears along the east Australian coast and fur seals *Arctocephalus* are a common sight on the shores of New Zealand and South America. The north Pacific also supports quite a variety of seals, the best known of which is the California sea lion *Zalophus*, an eared seal that has made a name for itself as a performer in captivity.

Other noteworthy examples are the monk seal *Monarchus* in the region of the Hawaiian Islands and, further north, the northern elephant seal *Mirounga*.

A marine mammal not related to either the seal or whale groups is the slow-moving browser of sea-weeds, the dugong or sea-cow *Dugong dugong* which occurs in the tropical and subtropical waters of

THE ATLANTIC OCEAN

The Atlantic, the second largest ocean, contains about 25 percent of all the water in the world ocean, and is about one-half the volume of the Pacific. The science of oceanography had its birth in the Atlantic, and concepts developed in the north Atlantic have been applied to other oceans with varying degrees of success. Generalizations have not always stood the test of time, since the north Atlantic differs in fundamental ways from other parts of the world ocean.

A UNIQUE STRUCTURE

The Atlantic may be visualized as an immense fiord, opening at its southern end into the Indian, Southern, and Pacific oceans, and essentially closed in the north. Its northern end is connected with the North Polar Sea, often called the Arctic Ocean although its volume is far too small for it to be considered as a true ocean. Exchanges between the North Polar Sea and the north Atlantic are minuscule compared with the exchanges that occur in the south.

Because of the fiord-like nature of the Atlantic, its northern and southern components are very different. The north Atlantic is the warmest and most saline of all the world's oceans, while the south Atlantic is more like the Indian and Pacific oceans, although it is slightly fresher and more saline than the oceanic average. The saltiness of the north Atlantic results from the outflow of the Mediterranean Sea, which discharges prodigious amounts of extremely saline water. This Mediterranean outflow dominates the eastern north Atlantic just as the Gulf Stream dominates the western north Atlantic.

Because the north Atlantic is not open at the north, its circulation is much more closed than that of the south Atlantic, and its currents are much stronger. In the south Atlantic, the warm Brazil Current is a weak analog of the powerful Gulf Stream, and the Benguela Current is a warmer analog of the Canary Current in the north Atlantic, transporting cool water and pelagic organisms from the Indian Ocean.

FAUNAL REGIONS

Based on collections of midwater fishes, the Atlantic Ocean has recently been divided into eight immense faunal regions by R.H. Backus and co-workers of the Woods Hole Oceanographic Institution. They tried to find physical oceanographic boundaries that corresponded with the limits of midwater fish distribution, thereby linking faunal patterns with specific water masses. From north to

FAUNAL REGIONS OF THE ATLANTIC OCEAN

south, these regions are: the Atlantic Subarctic, the North Atlantic Temperate, the North Atlantic Subtropical, the Gulf of Mexico, the Mauritanian Upwelling, the Atlantic Tropical, the South Atlantic Subtropical, and the South Atlantic Temperate. These faunal regions serve well to describe the distribution of a wide variety of pelagic vertebrates and invertebrates. In addition, they provide a good overview of the major water masses that make up the Atlantic Ocean.

The Atlantic Subarctic is a transition zone between the frigid waters of the North Polar Sea and the North Atlantic Temperate Region. It is characterized by extremely low diversity, with only a single species, *Benthosema glaciale*, making up the majority of the midwater fish biomass. Puffins and other seabirds nest along its coasts and on its islands; pilot and fin whales traverse its breadth on their oceanic journeys.

As one moves south toward the equatorial regions, biological diversity increases. The North Atlantic Temperate Region extends from the shores of Europe to the north-eastern part of the United States, and is highly productive, with marked seasonal changes in water temperature. The Grand Banks and North Sea, historically very important centers of fisheries activity, are located in this region, their harvests including pelagic herring, sardine, and anchovy, and demersal cod, flounder, and Atlantic perch.

The South Atlantic Temperate Region, extending to the south to the Antarctic Convergence, is open to the Indian and Southern oceans, and is bounded only on its western side by land (the east coast of Argentina). Since it is open to the Southern Ocean, species such as

The division of the Atlantic Ocean into faunal regions has provided a framework for the study of midwater fish and other marine animals.

David Doublet

This parrotfish Scarus vetula *has an unusual sleeping habit. At night, it positions itself in a crevice and spins a mucus bubble, which protects it against the sharp edges of the coral and possibly predators.*

Opposite. *Sluggish relatives of sharks, stingrays are named from the bony, poisonous sting embedded in their long, flexible tails. Many species inhabit sandy bottoms of shallow lagoons and coral reefs, sometimes in schools, and typically bury themselves in the sand with only the eyes showing. This pair are southern stingrays* Dasyatis americana, *lying in the shallows at Grand Cayman, Cayman Islands.*

whales, penguins, and seals that occur close to the Antarctic continent are also found in this region.

The North and South Atlantic Subtropical regions are areas of low primary production but high diversity. Warm water extends to great depths, and the surface waters are very clear but very low in the nutrients necessary for the growth of plants. This is because of the deep permanent thermocline, which prevents nutrient-rich deeper waters from reaching the surface. In the north Atlantic, the western portion of the Subtropical Region forms the Sargasso Sea. In the south Atlantic, a discrete "sea" does not form, because it is open to exchanges with the Indian and Southern oceans. Oceanic wanderers such as sharks and dolphins, whales and sea turtles, sunfish, ctenophores, and siphonophores cross the subtropical regions on their migration paths.

The Atlantic Tropical Region is centered on the thermal, rather than the geographical equator. The thermal equator is the region with the warmest surface waters, and runs diagonally in a wide swath from the Caribbean Sea to the coast of Angola. Its currents are very complex, and change direction seasonally. This part of the Atlantic is a region of extremely high faunal diversity.

In addition to these major Atlantic regions, Backus and his colleagues described two smaller regions that did not fit into the overall pattern. The Mauritanian Upwelling, off the coast of northwestern Africa, is extremely productive, and one species of midwater

David Doublet

A SEA BOUND BY OCEAN

JAMES C. KELLEY

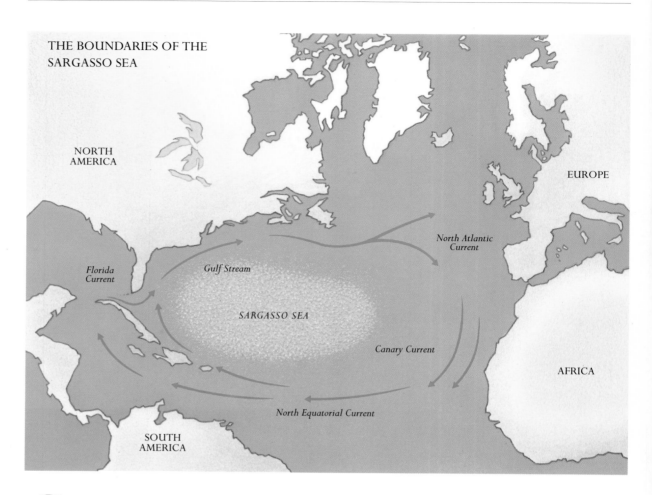

THE BOUNDARIES OF THE
SARGASSO SEA

NORTH
AMERICA

EUROPE

*North Atlantic
Current*

*Florida
Current*

Gulf Stream

SARGASSO SEA

Canary Current

AFRICA

North Equatorial Current

SOUTH
AMERICA

*C*hristopher Columbus encouraged his seamen to continue their westering voyage, bound for the New World, with these words from his log: "The Admiral says that on that day, and ever afterwards, they met with very temperate breezes, so that there was great pleasure in enjoying the mornings, nothing being wanted but the song of nightingales . . . Here they began to see many tufts of grass which were very green, and appeared to have been quite recently torn from the land."

Larry Lipsky/Tom Stack & Associates

The *Sargassum*, as it is now called, is a brown alga which looks very much like an attached marine plant, but which exists entirely in the pelagic, open-ocean environment. Like many plants, the *Sargassum* reproduces asexually, growing on one end and dying on the other. The *Sargassum* is only one of hundreds of life-forms unique to the Sargasso Sea, the only sea in the world bordered not by land but by the Atlantic Ocean. The great north Atlantic currents—the Gulf Stream, the North Atlantic Current, the Canary Current, and the North Equatorial Current—are a single closed-circulation cell called the North Atlantic Gyre.

The clockwise north Atlantic circulation has a number of special consequences. The Coriolis effect, a result of the Earth's spin, causes water in motion to move to the right in the Northern Hemisphere and to the left in the Southern Hemisphere. This causes anything that floats to become concentrated in the center of the gyre. The gyre is lens-shaped, varying in thickness from about 200 meters (650 feet) at the edges to perhaps 100 meters (330 feet) under the Gulf Stream and in the center, under the Sargasso Sea.

The central gyres, like the Sargasso Sea, have no local nutrient supply. The water is thus biologically poor, nearly devoid of life. It is the bluest, most translucent water in the ocean. Only jellyfish, salps, and other large animals are visible. In the last 20 years, the nutrient transport system into the Sargasso Sea has become much better understood. As the Gulf Stream moves north along the east coast of the United States, the stream meanders. Some of the meanders break off from the main current and become closed-circulation cells or "rings". Some of these cells trap nutrient-rich coastal water and transport it offshore, into the center of the Sargasso Sea. These rings have been tracked for several weeks by ships, and satellite data show that they may persist for months.

It is within the *Sargassum* itself that the real drama unfolds. A number of life-forms have co-evolved with the *Sargassum* and have taken on the yellow-brown coloration of the alga. They live their entire lives in its camouflage. Perhaps the most exotic is the sargassum fish *Histrio histrio* which crawls through the *Sargassum* with nearly prehensile fins. There are two common shrimps and a small crab, as well as a nudibranch. A tiny pipefish, a relative of the sea-horse, mimics the thin strands of the *Sargassum* perfectly. Several other cryptic animals— hydroids, bryozoans, corals, and other small marine invertebrates—also live only in the *Sargassum*.

The most famous inhabitants of the Sargasso Sea are not permanent residents. The eel *Anguilla*, a delicacy that graces European and American tables, spawns in the Sargasso Sea. The larvae spend from one to three

John & Gillian Lythgoe/Planet Earth Pictures

years in the gyre. They then migrate out of the oceanic waters into rivers in North America, western Europe, and the Mediterranean. When they are fully mature, they return to the sea and migrate back to the Sargasso to mate and spawn.

The Sargasso Sea is one of the most fabled parts of the ocean. The subject of many myths, it is really known only to professional seafarers who cross it on ship tracks from Europe to the Caribbean. It is the graveyard for hundreds of ships set adrift in the North Atlantic Gyre. More importantly, perhaps, it is among the most peaceful, least visited, and most beautiful parts of the world ocean. ●

Eels Anguilla, *recently developed from their larval stage into a transparent form, called elvers. It takes these young eels about two to three years to travel the distance from their hatching area in the Sargasso Sea to freshwater rivers and streams in Europe and North America.*

Robert F. Sisson

Within its swirling mass the Sargassum *supports a number of life-forms, including two common shrimps, a crab, a pipefish, a nudibranch, and several hydroids, bryozoans, and corals.*

Opposite. A well-camouflaged sargassum fish Histrio histrio *hides among the* Sargassum.

David Doublet

fish, *Lamadena pontifex*, is unique to it. On the other hand, while the Gulf of Mexico has no unique pelagic species, it has a fauna that is a mixture of temperate, subtropical, and tropical species.

THE SPECIAL NATURE OF THE ATLANTIC

The Atlantic Ocean started forming about 200 million years ago when the continents of North and South America began to separate from the continents of Europe and Africa, along a massive rift valley. Because the Atlantic is geologically so young, the bottom-living (benthic) animals that inhabit it are descendants of colonists from the other oceans and seas. Therefore, the diversity of the benthic fauna is not as great in the Atlantic as it is in the Indo-Pacific. For example, many genera of reef-building corals of the Pacific are absent altogether in the Caribbean.

Because the waters of the north Atlantic wash the shores of Europe and North America, it is not surprising that oceanography had its beginnings in the

north Atlantic. Early studies on the speed and direction of the Gulf Stream were prompted by economic forces. Those who could use the extra impetus provided by this "river in the ocean" could cut their transit times across the Atlantic, and thus increase their profits. Several nineteenth-century oceanographic expeditions had their focus in the Atlantic and, as a result, many of our present ideas about the nature of the world ocean are largely based on studies made in the Atlantic.

However, as we learn more about other parts of the world ocean, the atypical nature of the north Atlantic has become apparent. Processes such as the "spring bloom" (a rapid growth of phytoplankton as days become longer) are entirely lacking in other parts of the world ocean, such as the north Pacific. It is clear that a greater understanding of the world ocean as a whole is necessary before we can generalize with confidence.

RICHARD HARBISON

Some of the deep-ocean sharks rely on a sort of ram-jet ventilation to breathe, their own swimming motion forcing a steady stream of water through the gill chambers. More sedentary species, such as this Caribbean reef shark Carcharhinus perezi, *have more conventional respiratory mechanisms. With no need to keep moving in order to breathe, these sharks often rest on the bottom during the day, even "sleeping" in caves or crevices.*

Opposite. It is common in Caribbean waters for sponges of different species to live in close proximity. Here, red vase sponges and brown tube sponges share the same rock.

The scaly chromis Chromis lepidolepis, *a member of the family* Pomacentridae, *inhabits coral reefs in the Indo-Pacific region. Plankton feeders, these fish shelter at night in crevices and caves, and emerge at dawn to school and feed near the surface.*

Opposite. Coral reefs are to the sea what tropical rainforests are to the land: their diversity and structural complexity provide a multitude of microhabitats for a huge variety of fish species. Here a magnificent coral cod Cephalopholis *emerges from its home in a coral reef crevice in the Red Sea.*

THE INDIAN OCEAN

The Indian Ocean is less well studied than the other oceans. Life in the Indian Ocean is often considered as part of a wider region—the Indo-West Pacific—implying that there are many similarities between the Indian Ocean and the western Pacific. Many marine organisms have pelagic eggs or larvae which are dispersed by tides and currents. The passage of sea water across the north of Australia allows these organisms to pass between the two oceans either as larvae or as adults. The Indian Ocean covers a similar range of southern latitudes to the Pacific, and many species occur in both oceans.

The Indo-West Pacific region has the richest marine fauna in the world. The central area (roughly the Malay Archipelago, the Philippines, and New Guinea) has been suggested as the principal evolutionary center from which the entire Indo-Pacific has been populated. Diversities are higher there for corals, sponges, medusae, crustaceans, echinoderms, and fish. The fauna becomes progressively impoverished with distance from this center, but there are also many species which occur only in the Indian Ocean.

Open-ocean areas of the Indian Ocean are characterized by relatively low primary productivity. Phytoplankton and associated zooplankton densities are generally very low, and it is only adjacent to coasts that productivity is stimulated. Where plankton densities are more concentrated, either at particular depths or places, large filter-feeding organisms such as whale sharks and baleen whales gather. The smaller filter feeders and predators attract larger predators including sharks, dolphins, and whales.

INDIAN OCEAN FISH

There are between 3,000 and 4,000 species of tropical shore fish in the Indian Ocean, often forming bright colorful "clouds" around corals, but with many different roles to play in reef ecology. Many species have restricted distributions within the ocean, particularly in the Red Sea and the Arabian Gulf. Both these areas have been subjected to isolation because of sea-level changes and have thus been provided with opportunities for evolution of new species.

There are fewer species of fish on the continental slope and in deeper water, but they too exhibit a high degree of endemism: 58 percent of the species are

found only in the Indian Ocean. Pelagic fish are also
abundant, including needlefish which may be 1 meter
(3 feet) in length. Truly oceanic fish include sharks,
flying fish, tunas, marlins, and sunfish. Most of these
are widely distributed throughout the tropics and
migrate between feeding and breeding grounds. Tuna
fisheries are well developed in the Indian Ocean, and
the species caught include albacore, yellowfin, southern
bluefin, and billfish. These are very mobile and are
distributed across the ocean, with some migration
around the southern pole. Tuna migration and
physiological adaptations have been extensively
studied by scientists (see feature on page 58).

CORAL REEFS

Where the water temperature never falls below
18–20°C (64–68°F) in winter (between 30° latitude
north and 30° latitude south of the equator), coral
reefs are formed on shorelines throughout the Indian
Ocean. However, this pattern is modified where there
is freshwater runoff, particularly when it contains a
heavy sediment load off the land as occurs off the coast
of India. There, mudflats and mangrove swamps form
less picturesque, but no less important, habitats for
marine organisms.

Coral reefs in the Indian Ocean may be fringing
reefs, patch reefs, or atolls. Deep-water atolls occur
throughout the ocean. The longest fringing reef occurs
in the Red Sea, with a total length of 4,500 kilometers
(2,800 miles), but there are also extensive fringing
reefs off eastern Africa and along most suitable coasts
in the Indian Ocean. Patch reefs are extended areas of
comparatively shallow seabed with simple reefs. These
are common in the Red Sea but also occur elsewhere in
the Indian Ocean.

SEAGRASS BEDS

This region has some of the most diverse and abundant
seagrass beds in the world, particularly on the Western
Australian coastline. Seagrasses are descended from
terrestrial plants and, unlike algae or seaweeds, have
true roots and flower under water. They grow mainly
on sand in areas that have some protection from open-
ocean swell, in coastal lagoons, and behind reefs.
Seagrasses form a food supply for turtles and dugongs,
and some of the world's largest remaining dugong herds
occur in the Indian Ocean, especially in Shark Bay,
Western Australia. There are still dugongs elsewhere in
the Indian Ocean, for example in the Red Sea and the
Arabian Gulf, but these populations have been greatly
reduced by human impact.

Seagrass beds also act as a nursery for juvenile stages
of fish and prawns, providing a refuge from predators
and thus allowing these species to grow to adult size. In
some areas, such as the northern and western coasts of
Australia, lobsters also use seagrasses as their larders,

Peter Scoones/Planet Earth Pictures

foraging for small animals and debris that accumulate in seagrass meadows. Seagrasses slow the rate of water flow over them, helping to prevent erosion and sometimes building up extensive sandbanks. Over geological time these structures can accumulate more sand than even a coral reef. This sand comes partially from the water passing over the seagrass, but also from the calcareous animals and algae associated with the seagrass leaves. A good example of this is in Shark Bay, where sand structures over 10 meters (33 feet) thick and many kilometers in length have accumulated in about 5,000 years.

MANGROVE ZONES

Mangroves dominate the intertidal mudflat zones in tropical and some subtropical coasts of the Indian Ocean, and are associated with estuaries. There are many different types of mangroves, the most widespread being *Avicennia*, the red mangrove, and *Rhizophora*, the black mangrove, which are circumtropical. *Sonneratia* is common in the Indo-Pacific as the most seaward species, grading into

Avicennia and then inland into *Rhizophora*. Three species of *Rhizophora* are found in the Indian Ocean: *R. mucronata*, which is the commonest from east Africa to New Guinea; *R. stylosa*, which occurs from Malaysia to Australia; and *R. apiculata*, which has a similar distribution to *R. stylosa*, but extends to India. Mangroves are important sources of productivity, but they are extremely vulnerable to human impact. Reclamation work in heavily populated areas of the Indian Ocean has resulted in almost complete removal of mangroves from some areas, such as around Singapore and elsewhere in developed Southeast Asia.

DIANA WALKER

The warm, sheltered, and shallow waters of Shark Bay on the coast of Western Australia support the most extensive seagrass beds in the world, and provide perhaps the single most important refuge for dugongs left anywhere.

The dugong shares its environment with another large marine herbivore, the green turtle Chelonia mydas. *Unlike dugongs, which live permanently in groups, the green turtle is a solitary animal that congregates only to breed.*

Blue swimmer-crabs **Portunus pelagicus** *are carnivorous, and spend most of their time lying in wait for prey while buried in the sandy bottom with only their eyes showing.*

AN UNDERWATER MEADOW

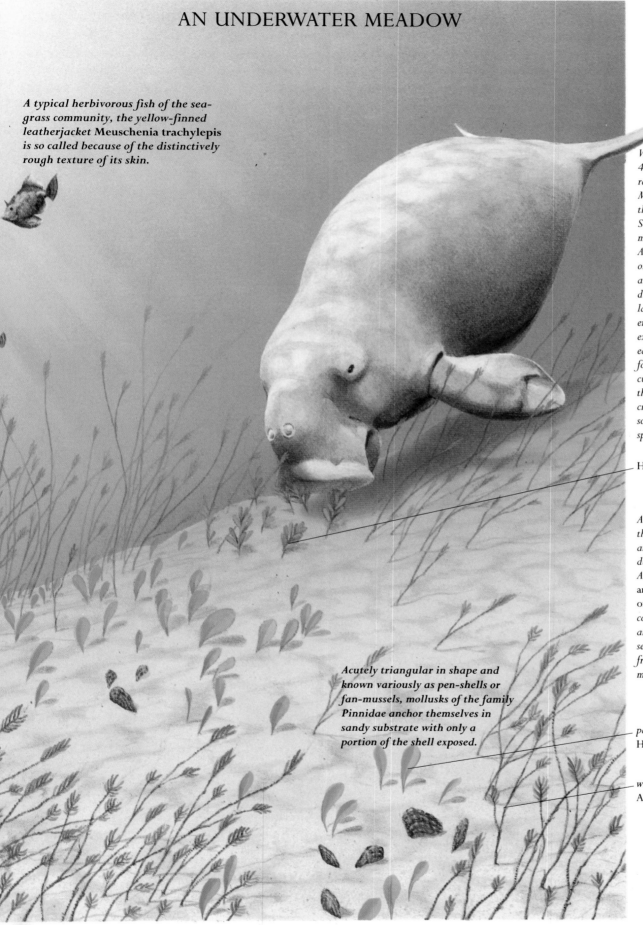

A typical herbivorous fish of the sea-grass community, the yellow-finned leatherjacket Meuschenia trachylepis *is so called because of the distinctively rough texture of its skin.*

With a distribution spanning some 43 nations in the Indo-Pacific region from New Caledonia to Mozambique, the dugong is one of the few surviving members of the Sirenia order; three species of manatees in the Caribbean, Amazonia, and West Africa are the only others. With few predators and few competitors for food, dugongs grow large, move languidly, and expend little energy—only about a third of that expended by a land mammal of equivalent size. Frequently hunted for food by humans in traditional cultures, and a common victim of the propeller blades of pleasure craft in more industrialized societies, the dugong is a declining species in many areas.

Halophila spinulosa

A plant community of at least three species of seagrass provides an important sanctuary for dugongs in Shark Bay, Western Australia. Wireweed Amphibolis antarctica, *paddleweed* Halophila oralis, *and* H. spinulosa *constitute their main diet. These are true flowering plants, not seaweeds (algae), and differ from land plants only in their marine habitat.*

Acutely triangular in shape and known variously as pen-shells or fan-mussels, mollusks of the family Pinnidae anchor themselves in sandy substrate with only a portion of the shell exposed.

paddleweed
Halophila oralis

wireweed
Amphibolis antarctica

The Southern Ocean

The Southern Ocean completely encircles the Antarctic continent, its currents flowing in a predominantly west-to-east direction. Its northern boundary is the Antarctic Convergence at about 45–55°S where the ocean meets and flows under the warmer waters of the north. This is a region of strong winds, turbulent seas, and an almost continuous cloud cover. Southward from the convergence the temperature of the ocean becomes progressively cooler and ranges from 8 to 11°C (46–52°F) in the north to minus 1.8°C (29°F) at the coast of Antarctica. In winter, sea ice forms first around the coast of Antarctica and extends to cover more than 50 percent of the Southern Ocean.

The seemingly inhospitable nature of the Southern Ocean belies its productivity, which is probably as high on an annual basis as the oceans in more temperate climates. Plants and animals are tolerant of the cold and have features that fit them to their environment. The blanket of ice calms the surface of the ocean. Its undersurface provides a habitat for zooplankton and allows sea-ice algae to develop; and the upper surface supports breeding seals and emperor penguins.

At the base of the food web are primary-producing organisms—the microscopic algae of the phytoplankton and the sea-ice microbial communities. In the presence of light, these organisms produce food in the form of carbohydrates from carbon dioxide and water. Because the amount of daylight varies considerably, and at the coast of Antarctica ranges from almost continuous darkness in winter to continuous daylight in mid-summer, primary production is markedly seasonal. It is highest in the coastal zone where upwelling brings

Opposite, top. Two Antarctic dwellers of the phylum Cnidaria: a sea-anemone Utricinopsis *and a hydroid* Primonella.

Opposite, center. Although only a few centimeters in length, krill Euphausia superba *swarm in the Southern Ocean in such numbers that they may be the dominant species on Earth in terms of sheer biomass.*

Opposite, bottom. Only several species of starfish inhabit the frigid waters of Antarctica.

The most numerous of all pinnipeds, crabeater seals Lobodon carcino-phagus *are residents of the pack ice fringing the Antarctic continent.*

David Rootes/Planet Earth Pictures

nutrients to the surface. In spring the biomass of phytoplankton increases rapidly, to be consumed by the microzooplankton (mostly protozoans and zooplankton larvae) and the slightly larger zooplankton.

There are three major groups of herbivorous zooplankton: the euphausids, the salps, and the amphipods, all of which usually form dense aggregations. Of these, the euphausid *Euphausia superba* or krill is the most important as it provides the major link in the transfer of energy from the primary producers to the larger carnivorous organisms including the baleen whales, seals, seabirds, fish, and squid. Humans are also predators of krill and harvest up to 500,000 tonnes a year.

FISH AND SQUID

The fish of the Southern Ocean are predominantly bottom-dwelling species inhabiting the coastal shelves of Antarctica and the subantarctic islands. Of the world's 20,000 species of fish, only 120 occur in the Southern Ocean—but most of these are found nowhere else. About 75 percent of the fish species belong to the group of Antarctic cods, and many live in an environment of minus 1.8°C (29°F). Their body fluids contain anti-freeze substances. One group of "icefish" (Chaenechthyds) have no hemoglobin, the substance that carries oxygen in the blood. Instead they absorb oxygen directly into their tissues, principally through the gills. Many of these species have been harvested commercially and some have been substantially overfished, particularly around the islands of South Georgia.

Squid are important in the food web of the Southern Ocean as they appear in the diet of many species of albatrosses and penguins and in particular the sperm whale. However, as they are difficult to catch, little is known of their biology.

SEALS AND SEABIRDS

Four species of seals breed on the Antarctic pack ice. The Weddell seal *Leptonychotes weddelli* prefers the stronger ice nearest the Antarctic coast and maintains ice holes that allow it to move between the air it must breathe and the sea where it feeds, principally on fish. With a population of 2 million or more, the crabeater seal *Lobodon carcinophagus* is the most numerous of all the world's seals. It inhabits the inner pack-ice zone in the vicinity of its major food source, the Antarctic krill.

Leopard seals *Hydrurga leptonyx* and Ross seals *Ommatophoca rossii* breed in the outer pack-ice zone on ice that is broken into floes, allowing easier access to the ocean. Both feed on krill; the leopard seal also feeds on penguins and has been observed to attack the pups of the crabeater seal. The Antarctic fur seal *Arctocephalus gazella*, which breeds on the subantarctic islands of the Indian and Atlantic oceans, is now recovering from heavy sealing and is expanding its range from South

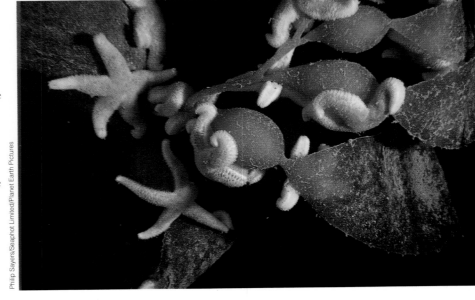

The icefish Chaenocephalus aceratus *represents an extra-ordinary group of fish that have no hemoglobin in their blood-stream, hence their pallid appearance. Glycerine-like compounds in and between the body cells function like a natural anti-freeze, preventing tissues from freezing in the frigid waters surrounding Antarctica.*

Georgia down to the Antarctic Peninsula where, like leopard seals and Ross seals, it feeds on Antarctic krill.

Some 35 species of seabirds breed on the Antarctic continent and subantarctic islands. Typical are the penguins, specialized for pursuing their prey under water in a cold marine environment, and the albatrosses, equally specialized marine species able to ride the Antarctic gales with seemingly effortless ease. Other species include the petrels, prions and storm petrels, and members of the gull family (skuas, a gull, and a tern). The Adelie penguin is the most important of the seabirds, both numerically and in its consumption of krill which in total approaches that consumed by the crabeater seals and about half that of the baleen whales. It inhabits the coastline of Antarctica, nesting in large colonies.

WHALES AND THEIR FUTURE

Seven species of baleen whales, which feed predom-inantly on krill, and eight species of toothed whales, which feed on squid and fish, and some on penguins, seals, and other whales, flourish in the Southern Ocean. The whaling industry focused on the largest of the baleen whales—blue, sei, fin, and humpback—and reduced their populations to very low levels. Even with the continued prohibition on whaling of these species, recovery of the populations to former levels is doubtful, and the very survival of the blue whale is at risk. The toothed sperm whale is also at risk and is now protected.

While they are in Antarctic waters during the summer months, the present stock of baleen whales consumes some 40 million tonnes of krill, compared with about 150 million tonnes before whaling. The difference, termed the "krill surplus", may not necessarily be available to the large whales to rebuild their numbers. The penguins, seals, and smaller species of whales, which have a shorter generation time, are able to build up their populations more rapidly than the great whales, and thus become more powerful competitors. Currently, there is, in fact, an increase in numbers of penguins and Antarctic fur seals.

CONSERVATION OF LIVING RESOURCES

An international agreement, the Convention for the Conservation of Antarctic Marine Living Resources, has been in force since 1982. Its prime purpose is to ensure rational use of the living resources of the Southern Ocean and it is attempting to manage the fisheries in an ecosystem context. Though progress is slow, the fundamental issues are set in a conservation standard which requires that all dependent and related species, and the recovery of depleted stocks (such as whales), are taken into account in the management of the fisheries. This convention applies to the whole Southern Ocean and is thus unique in the world. ■

KNOWLES KERRY

Opposite. *Rockhoppers* Eudyptes chrysocome *coming ashore to visit their nesting colony, New Island, Falklands. On land, most penguins walk—or at least waddle—but rockhoppers, widespread in the Southern Ocean, are named for their persistent habit of hopping along, like a person in a sack-race.*

6 THE OCEAN DEPTHS

ROBERT R. HESSLER, SCOTT C. FRANCE, AND MICHEL A. BOUDRIAS

Because of its inaccessibility and extreme conditions, the deep sea has always been regarded as a mysterious and forbidding place. It has at once been considered as home to fantastic monsters and as lifeless as space. Over the last century technological developments have allowed scientists to probe deep below the waves and into the abyss. Surprisingly, far from being an uninhabited wasteland, the deep sea has proven to be home to a diverse collection of fish and invertebrates that are adapted to living within this environment of extremes.

The bottom-dwelling deep-sea prawn Nephropsis atlantica *is a deposit feeder. The well-developed setae on its front limbs assist it to detect food particles on the sediment surface.*

These squid Pyroteuthis margaritifera *inhabit the ocean depths. Relatives of snails and shellfish, squids swim by jet propulsion, violently expelling water from the body cavity, and have highly developed sensory systems, including visual equipment nearly as sophisticated as that of mammals and birds.*

THE DEEP SEA

The deep sea is that area of the ocean below 200 meters (650 feet). This is the depth where, in the clearest oceanic waters, so much light is absorbed that photosynthesis does not support phytoplankton. Why is this environment so different from the one we inhabit? The most important factors are pressure, temperature, and darkness.

On land we are accustomed to living under one atmosphere of pressure, the weight of the entire air column above us. Water is much heavier than air; in water, for each 10 meters (33 feet) of depth, the pressure increases by another atmosphere. Therefore, in the deepest parts of the ocean the pressure is over 1,000 atmospheres.

However, while this pressure certainly has an effect on the physiology of living organisms, it does not preclude life. Under high pressure water is virtually incompressible, but gas-filled spaces can be crushed. Most deep-sea animals have no air spaces, such as lungs and swim bladders, in their bodies. Those deep-living fish that have a gaseous swim bladder are able to control its internal pressure, unless they change depth rapidly. What appears forbidding to humans need not be a hindrance to other life-forms.

A CONSTANT ENVIRONMENT

Unlike the surface of the ocean, where tropical water temperatures can be above 30°C (90°F) and polar temperatures dip below freezing, the deep sea is cold throughout, ranging from minus 1 to 2°C (30–34°F). The variety of surface-water temperatures creates many habitats, to which different animals have adapted. Thus, shallow-water communities differ from the polar seas to the tropics. In contrast, the constancy of deep-sea temperatures erases the boundaries between latitudinal zones: the deep sea does not get warmer from the poles to the equator, but is rather a single, cold bath. As a result, deep-sea animals are free to live throughout the abyss without adapting to different climates.

This might seem surprising in view of the vastness of the environment. The deep sea extends over 341 million square kilometers (132 million square miles) —67 percent of the globe—at an average depth of 3,800 meters (12,500 feet). Yet there are only three basic habitats: the water column, or pelagic zone; the soft bottoms of accumulated terrestrial particles and planktonic remains; and the rocky bottoms of undersea mountain ranges, volcanoes, and seamounts. Sampling of these habitats reveals little relationship between latitude and the distribution of organisms. Populations of closely related animals are found from one end of the deep ocean to the other. Species vary from place to place, but the genera they belong to are found throughout. This evidence indicates that, despite its vastness, the entire deep sea is relatively homogeneous.

THE ABSENCE OF LIGHT

The factor of greatest biological importance in this environment is the absence of light. Light is fundamental to all life. Sunlight is the energy source which powers the food web upon which all animals, including humans, depend. Plants—in the ocean the microscopic phytoplankton and seaweeds—are at the center of this web, using solar energy to convert water and carbon dioxide into carbohydrates through the process called photosynthesis. Carbohydrates are the "bricks" of nature, the macromolecules that organisms need to grow, reproduce, and maintain themselves. The plants, in turn, are eaten by herbivorous animals, which are themselves eaten by carnivorous animals, and thus the entire food web owes its existence to solar energy.

Clearly, the absence of light is a critical feature of the deep-sea environment. Without light there can be no photosynthesis and thus no plants, no local production of food. Therefore, all the creatures that live in this vast expanse are dependent upon food which is ultimately produced in the thin layer of sunlit surface water. But only about 1 percent of the food produced at the surface reaches the bottom of the deep sea! A crude analogy might be one in which only the inhabitants of the top floor of a large apartment building produce food; whatever they do not eat themselves trickles down to the floors below.

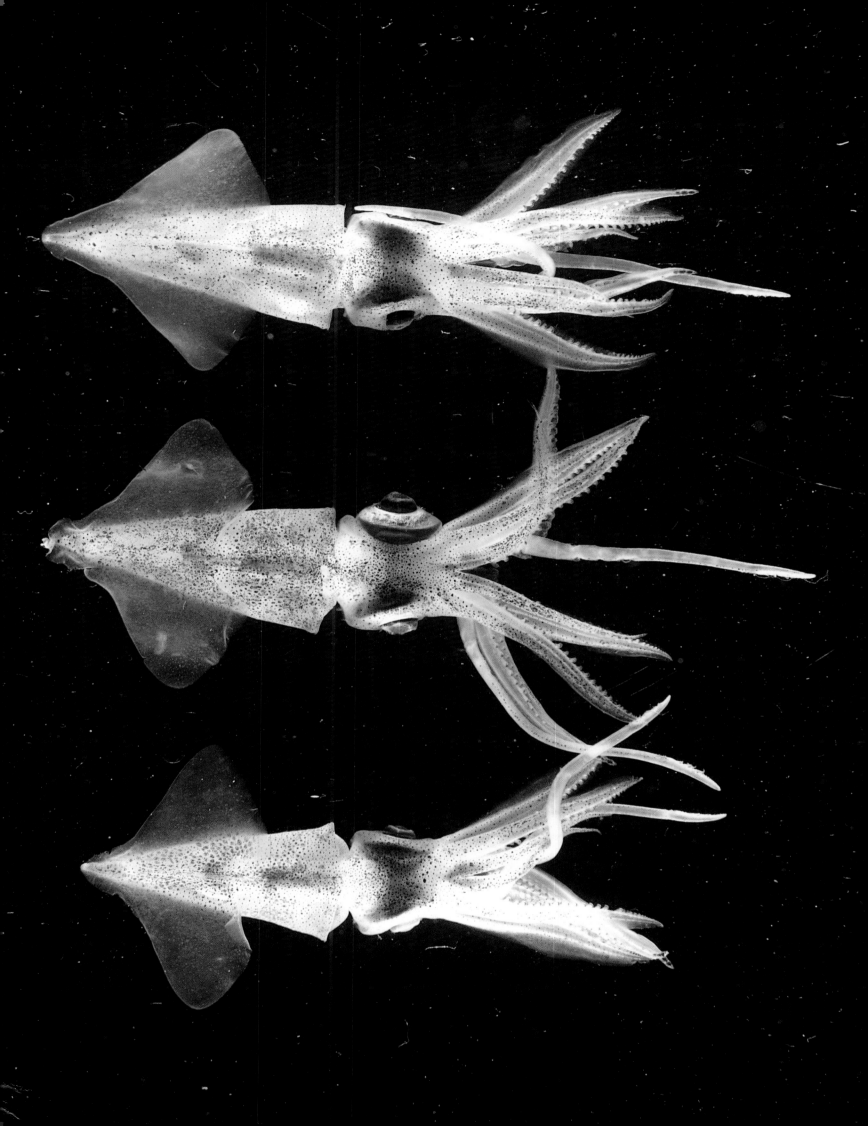

FROM THE SURFACE TO THE OCEAN DEPTHS

Below 1,000 meters (3,300 feet), the deep sea is dark and, to most life-forms, forbidding.
But some creatures have adapted to this habitat.

EPIPELAGIC
(0–200 meters / 0–650 feet)

1   cod
2   green turtle
3   mackerel
4   flying fish
5   Portuguese man-o'-war
6   dolphin
7   bluefin tuna
8   prawn
9   sperm whale
10   squid
11   shark
12   swordfish

MESOPELAGIC
(200–1,000 meters /
650–3,300 feet)

13   hatchetfish
14   dragonfish
15   lanternfish
16   viperfish
17   demersal octopus

BATHYPELAGIC
(1,000–4,000 meters /
3,300–13,000 feet)

18   gulper eel
19   anglerfish
20   deep-sea eel
21   bivalve
22   anglerfish
23   whalefish
24   cusk eel
25   rattail
26   brittlestar
27   crinoid
28   short-armed starfish
29   glass sponge
30   tripod fish
31   lamp shell

Residents of each successive floor eat their portion of whatever they can catch. Little remains by the time the food reaches the ground floor.

Life in the deep sea is not only dictated by darkness itself, but is strongly affected by the resulting low food levels and the need to utilize energy effectively. For example, in order to reduce the amount of energy spent maintaining position in the water column, animals have found ways of floating passively: by replacing heavy bone with lighter material such as cartilage; by incorporating fat, which is lighter than water, in body tissues; or by increasing the body's water content. On the whole, life in the bathypelagos proceeds at a languid pace with most animals casually swimming or drifting, simply waiting for food to come to them.

THE TWILIGHT ZONE

Sunlight penetrates 1,000 meters (3,300 feet) into the ocean, though it supports photosynthesis only in the upper 200 meters (650 feet). Between 200 and 1,000 meters is a twilight zone, where only blue light remains. By far the greatest volume of the deep sea is below that, enveloped in a world of perpetual darkness where the only light perceived is biologically produced.

Many midwater organisms use this biologically produced light—bioluminescence—in a variety of ways. Some animals flash unique patterns to identify and attract mates in the dark. In the twilight zone, several kinds of fish and shrimp have bioluminescent organs on the lower half of their bodies which act as a camouflage mechanism. These animals mimic the wavelength and intensity of sunlight filtering down from the surface, thus vanishing from the view of upward-looking predators searching for the silhouette of prey.

In the deeper zones of the deep sea, large gelatinous organisms (ctenophores, salps, medusae) use bioluminescence as a startling mechanism. When touched, they produce a burst of light that may temporarily blind or distract their predators. Some copepods expel bioluminescent clouds to confuse their attackers as they escape. Some of the bottom-dwelling rattails and other species have internal light organs, and the light shines through scaleless windows on the belly or through the body wall. Similarly, the light in the lures of deep-sea anglerfishes comes from a colony of bacteria within the lure.

DETECTING PREY

Bioluminescence may also be a hunter's tool. Many deep-sea predators expend little energy searching for prey, but instead find ways to attract them. The anglerfish, the most notorious of this genre, waves a bioluminescent lure at the top of its head to entice smaller fish, then captures them with a rapid gulp of its large, armored mouth. It has been suggested that krill emit light from the eye, like a flashlight, to illuminate prey which have struck its antennae.

Agence Nature/NHPA

Body pigmentation may also be used as camouflage against predators. Below 500 meters (1,650 feet) most deep-sea fish are black, while crustaceans are deep red. In limited blue light, this renders them effectively invisible. How, then, do predators detect their "invisible" prey? Many have evolved alternate sensory modes. A number of fish species detect sound waves generated by motion. Sensory hairs in internal canals or on the body surface respond to small changes in water pressure around the animal. Some shrimp have antennae, several times as long as their body, which spread out in the water waiting to be touched.

Many of these animals are also capable of detecting low concentrations of odor. They can smell food, or even potential mates, from several meters away, and then follow the odor trail wafting from its source. These strategies are also employed by organisms that live on the bottom of the ocean.

BOTTOM DWELLERS

Bottom-living animals differ from water column inhabitants in many ways. For unknown reasons, bottom dwellers do not use bioluminescence. Hence vision is unimportant, and most benthic creatures either have no eyes or are blind. Bottom dwellers have little or no pigmentation; most are white. The occasional splash of color—orange-tinted foraminifera, reddish brittlestars, yellow sea-lilies—may result from ingested food or be an incidental byproduct of some physiological process.

The animals that live on the bottom of the deep sea generally belong to the same taxa (groups) one finds in

Colossendeis colossea is a deep-water species of marine spider found at depths of 5,000 meters (16,500 feet). Deep-water spiders have no eyes and feed by using their proboscis to extract juices from worms and other soft-bodied invertebrates.

IFREMER

Norbert Wu

The modified fins of the tripod fish Benthosaurus *allow it to "sit and wait" above the bottom in the faster-moving currents where it is better able to detect odors from food.*

Above, center. The gigantic pelican-like jaws of the gulper eel Eurypharynx *enable this fish to swallow prey that is up to twice its own size. As with most bathypelagic vertebrates, the gulper's teeth and skeleton are flimsy.*

shallow water: for example, annelids (bristleworms), mollusks (snails and bivalves), and crustaceans (such as shrimp, amphipods, and isopods). But at the lower levels of taxonomic classification—species, genus, and more rarely family—the animals are usually different. A species that can live in shallow water could not survive in the deep sea, and vice versa. Interestingly, most deep-sea animals are small: a typical deep-sea isopod, for example, is smaller than one from shallow water.

OBTAINING FOOD IN THE OCEAN DEPTHS

The single most important factor that molds the way deep-sea animals live is scarcity of food. In the depths of the sea, nutrition comes as a weak rain of particles from the surface, ranging in size from small zoo-plankton and fecal pellets to plant debris and large fish or marine mammal carcasses. Bottom-living animals feed on this food in three basic ways: they pick up deposited detritus from the bottom; filter suspended food out of the water; or eat other animals. All three feeding types live in the deep sea, but in different

proportions, depending on the currents and whether the bottom comprises soft sediment or rock.

Deposit feeders are the most common. These animals either live a sedentary life, often inhabiting tubes or burrows, or roam over the mud leaving tracks in their wake. Some, such as holothurians (sea cucumbers), ingest the surface layer of mud and digest what little food it contains. Others, including most of the crustaceans, carefully select only the nutritious particles. The most diverse group of bottom dwellers, the polychaetes or bristleworms, include some with long tentacles which extend from burrows to sweep the bottom for food particles.

Suspension feeders obtain food either by filtering large volumes of water or by spreading a net of arms or tentacles to capture their meal. This way of life is difficult in the deep sea, where few particles are in suspension and currents carrying particles toward animals are typically slow, moving approximately 2 centimeters (1 inch) per second. Many suspension

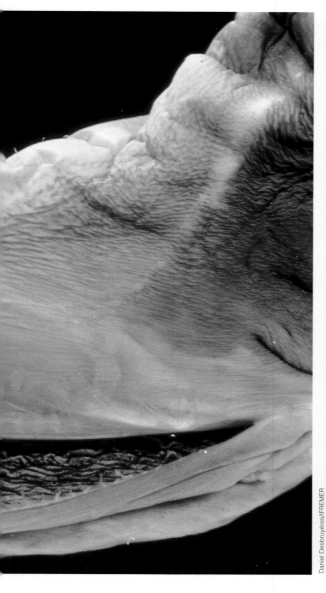

Daniel Desbruyères/IFREMER

feeders also need hard surfaces for attachment. For these reasons, suspension feeders are less common than deposit feeders in the deep sea, particularly on soft bottoms.

Because suspension feeders depend on current flow to bring their food, they benefit by being exposed to faster moving water. As current speed is greater slightly above the bottom, many suspension feeders have evolved ways to gain additional height, thus placing themselves within the best zone of suspended food. For example, glass sponges grow on long, delicate stalks; deep-sea corals can be 2–3 meters (6–10 feet) high, with complex branching patterns reminiscent of small trees; and many echinoderms, such as the brittlestars and brisingid starfish, climb the stalks of other animals.

On hard bottoms, topographic relief may intensify the speed of passing currents. It is here that suspension feeders are most likely to be concentrated. Even on small scales, such as the top surface of manganese

nodules, only centimeters above the bottom, small suspension feeders can dominate, taking advantage of slightly enhanced water movement.

Strict carnivores are relatively uncommon on the deep-sea bottom, perhaps because there is too little for them to hunt. However, many fish and crustaceans that swim above the bottom are specialized for consuming carrion. Much like vultures, they are able to detect infrequent food falls and descend on them within hours of their arrival on the bottom. Since they must often go for long periods without food, they are effective in gorging themselves when the opportunity arises. Thousands of amphipods have been recorded, even in trenches at depths up to 10,000 meters (33,000 feet). They attack fish carcasses like maggots and within hours devour them to the bone.

As a result of the lack of food, the abundance of life in the deep sea is low, hundreds of times lower than in shallow water. For decades scientists studying soft bottoms thought that because of this low abundance,

The hard-bottom substrate of the mid-ocean ridges is dominated by suspension-feeders such as anemones and stalked glass-sponges.

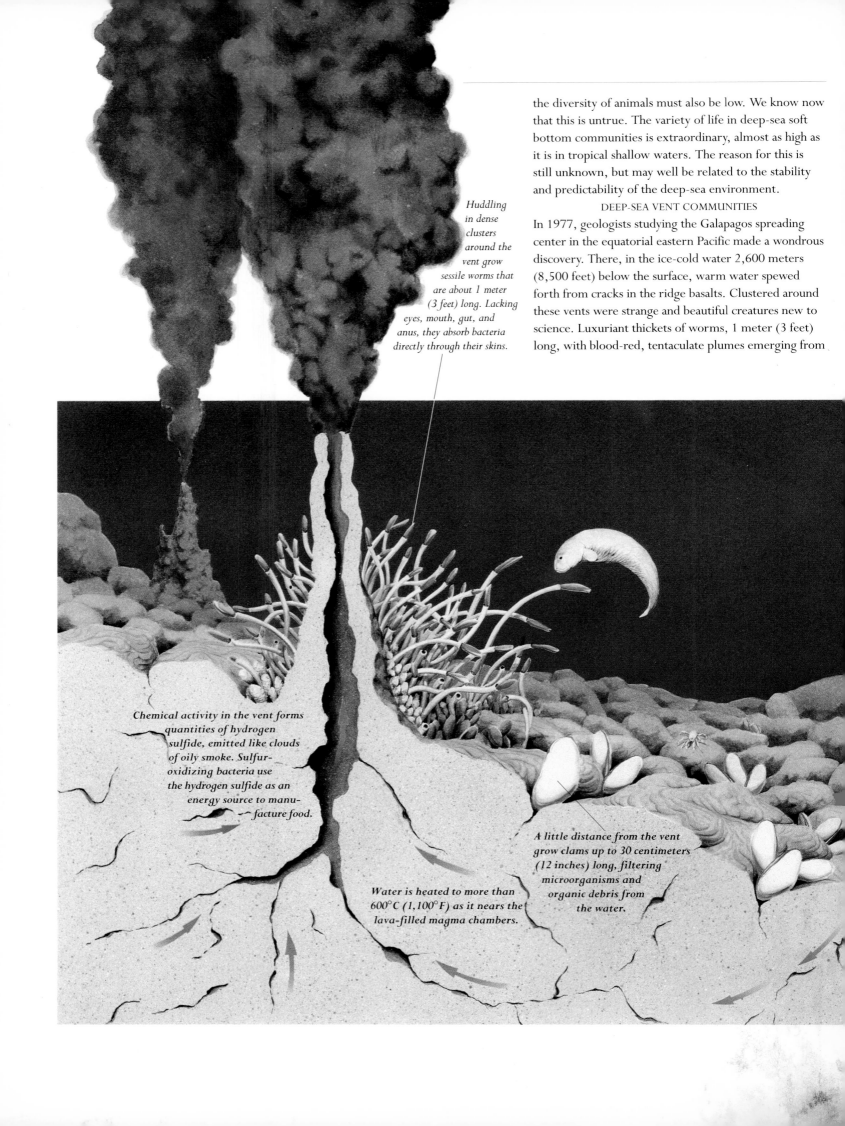

the diversity of animals must also be low. We know now that this is untrue. The variety of life in deep-sea soft bottom communities is extraordinary, almost as high as it is in tropical shallow waters. The reason for this is still unknown, but may well be related to the stability and predictability of the deep-sea environment.

DEEP-SEA VENT COMMUNITIES

In 1977, geologists studying the Galapagos spreading center in the equatorial eastern Pacific made a wondrous discovery. There, in the ice-cold water 2,600 meters (8,500 feet) below the surface, warm water spewed forth from cracks in the ridge basalts. Clustered around these vents were strange and beautiful creatures new to science. Luxuriant thickets of worms, 1 meter (3 feet) long, with blood-red, tentaculate plumes emerging from

Huddling in dense clusters around the vent grow sessile worms that are about 1 meter (3 feet) long. Lacking eyes, mouth, gut, and anus, they absorb bacteria directly through their skins.

Chemical activity in the vent forms quantities of hydrogen sulfide, emitted like clouds of oily smoke. Sulfur-oxidizing bacteria use the hydrogen sulfide as an energy source to manu-facture food.

A little distance from the vent grow clams up to 30 centimeters (12 inches) long, filtering microorganisms and organic debris from the water.

Water is heated to more than 600°C (1,100°F) as it nears the lava-filled magma chambers.

white tubes, gave the impression of a giant rose garden. Enormous 30-centimeter (12-inch) clams and piles of large mussels lay among the worm thickets. Crabs and shrimps by the dozen clambered over the sedentary fauna in search of food; tiny bristleworms projected their tentacles from tubes on the rock; small anemones carpeted the hard sea floor; and strange spaghetti-like acornworms and gelatinous "dandelions" (siphonophores) sat delicately around the edges of the vent field.

The discovery of this deep-sea hydrothermal vent community opened the door to an exciting period of research that has revealed vents on spreading center ridges throughout the world ocean.

The vent ecosystem is an anomaly in the deep sea: an "oasis" where a well-fed, teeming community thrives in an otherwise food-poor habitat. How can these animals grow so large and in such abundance in the deep-sea "desert"? The answer lies with the chemical content of the emerging hot vent water and the bacteria that take advantage of it. At these undersea vents, water seeps down through cracks in the rock and is superheated to more than 600°C (1,100°F) as it nears lava-filled magma chambers.

The composition of dissolved constituents in the hot water is altered via reactions with the rock. Biologically, the most important change is the addition of hydrogen sulfide, which sulfur-oxidizing bacteria use as an energy source to manufacture carbohydrates through chemosynthesis. This process is similar to photosynthesis, but differs significantly in that it does not rely on sunlight,

OASES OF THE DEEP

Several little-known fish species inhabit the fringes of the community.

Free-living scavengers, galatheid and brachyuran crabs scavenge among the clams.

Cold water percolates down through the sea floor.

In places on the deep ocean floor, several kilometers below the surface, volcanic activity allows water to percolate down beneath the sea floor to be superheated and returned as a fountain of hot water gushing forth into the ice-cold darkness of the abyss. These vents form oases in the deep sea, supporting some of the most bizarre ecosystems to be found anywhere.

Norbert Wu

The deep-sea anglerfish Melanocetus johnsoni *uses a modified dorsal spine as a built-in "fishing pole". The spine is topped with bioluminescent tissue.*

Opposite. *The fine hairs covering* Caulophryne jordani*'s body are part of the nervous system of the anglerfish and assist it in detecting prey.*

provided with an ideal incubation chamber; the worms benefit by receiving nutrition from the bacteria. Symbiotic relationships with chemosynthetic bacteria have also been noted in vent bivalves and snails, which harbor the bacteria in their gills.

THE VENT PARADOX

Living in a community based on hydrogen sulfide is not without problems. Hydrogen sulfide is toxic to one of the basic metabolic processes of all animals, cellular respiration, and poisons the flow of oxygen in the blood by pre-empting the oxygen-binding site of hemoglobin. Thus we have a paradox: a compound which is a beneficial energy source is poisonous to the organisms that rely on it!

But vent organisms have evolved ways of coping with this toxicity. The tubeworm binds the sulfide to a second site on the hemoglobin molecule. Thus one molecule can transport both oxygen and sulfide in the blood. The vent clam has evolved an entirely new molecule whose sole function is to bind and transport sulfide, thus leaving hemoglobin with the sole task of transporting oxygen. In the crabs, which have no symbiotic bacteria, sulfide is converted to a less toxic compound in the liver. The fact that alternative mechanisms have evolved to overcome sulfide toxicity is a good indication that the vent habitat, with all its food, is a preferred place to live and that vents have been around long enough for this evolution to occur.

LOCATING NEW HABITATS

Even though the hydrothermal vent habitat has been in existence for millions of years, individual sites are much more ephemeral. In the process of heating sea water, the magma chamber slowly cools off, with the result that circulation slows and the chemical changes diminish. Alternatively, the plumbing in the rock can be closed off by earthquakes or chemical precipitation. In either case, the vent dies, and the community dies with it. To avoid extinction, it is imperative that vent species be able to disperse to new vent habitats. For most species, the adults are either attached to the rock or too clumsy to be able to walk very far. They must rely on dispersal of their larvae. Scientists have yet to determine how offspring locate new vent sites. But, since food is plentiful, organisms can produce many young. These may drift in the water until, with luck, they find the needle in the haystack, another hydro-thermal vent. Most will not, and they will die.

Much remains to be studied in this fantastic, newly discovered habitat. How does the community change through time? What are the similarities and differences between vent communities in different oceans? What new kinds of animals are waiting to be discovered? The hydrothermal vents hold secrets which will keep deep-sea scientists busy for many years to come. ■

the fundamental source of energy on Earth. At vent eco-systems, sunlight is supplanted by geothermal energy.

Scientists studying the vent water found abundant chemosynthetic bacteria living in the hot water and growing on the rocks around the vent openings. More interestingly, when biologists closely examined the vent animals they found that several species had a mutually beneficial relationship, or symbiosis, with chemosyn-thetic bacteria living within their tissues. For example, the giant vent tubeworm has no mouth or intestinal tract. The trunk of its body contains a large organ packed with sulfur-oxidizing bacteria. The worm extends its plume into the warm water to absorb hydrogen sulfide and other necessary inorganic chemicals which the blood carries to the bacteria. The bacteria benefit by being

FISH OF THE DEEP SEA

JOHN R. PAXTON

There are two basic kinds of deep-sea fish: the pelagic fish of the midwaters and the bottom-dwelling benthic fish. Midwater fish between 200 and 1,000 meters (650–3,300 feet) live in the twilight or mesopelagic zone where the remaining light is slowly absorbed by sea water. Those midwater fish below 1,000 meters (3,300 feet) live in the total darkness of the true deep sea, the bathypelagic and abyssopelagic zones.

Peter David/Planet Earth Pictures

Two characteristic inhabitants of the deep sea, an environment so alien that only Latin names attach to most species: Pseudoscopelus (above) and Xenolepidichthys (below). The latter has a gut that is set along an axis at right angles to its body.

Peter David/Planet Earth Pictures

Although the deep sea is the world's largest habitat, only some 2,500 species—about 10 percent of all fish—live there. Of these, about 1,500 species are bottom dwelling, 850 pelagic species live in the twilight zone, and 300 midwater species live below 1,000 meters (3,300 feet). Some species move between the zones. Rattails are the dominant family of bottom fish, while lanternfish, lightfish, and dragonfish predominate in the twilight zone. Anglerfish and whalefish are the most common bathypelagic fish below 1,000 meters (3,300 feet).

Most deep-sea fish belong to primitive groups, like sharks, eels, and the less advanced bony fish families. Their environment is harsh, with low levels of food, light, and temperature. Presumably because of competition from the more advanced groups of bony fish, deep-sea fish have successfully adapted to this habitat. The lack of light and low level of food have been the most important evolutionary forces molding their striking adaptations.

VERTICAL MIGRATION

Many twilight-zone fish migrate from the daytime depths of 500 to 1,000 meters (1,650–3,300 feet) to the upper 200 meters (650 feet) at night, to feed in the rich surface waters. During daylight hours these fish would be easy prey for the surface predators in the shallows. But at night their adaptations to the dark deep sea make them equally at home in shallower water. A number of bottom-dwelling fish also rise from the bottom, particularly at night, but the distances they travel are usually much less. There is no evidence that the midwater fish living below 1,000 meters (3,300 feet) undertake a vertical migration. With no light there would be no daily cues for such movement.

FOOD AND FEEDING

As all food is produced in the surface waters, the amount of food decreases with increasing depth. A striking feature of most deep-sea fish is a very large mouth, often with large or specialized teeth. Anglerfish, for instance, have long, slender teeth that bend into the mouth as prey enter, but lock upright if prey attempt to go backward out of the mouth. Other fish and crustaceans are the food of most pelagic deep-sea fish, and a number of species, like anglerfish and swallowers, eat fish longer than their own body. With sparse food and slow digestion due to low temperatures, meals may be infrequent.

SENSE ORGANS

The eyes of deep-sea fish are variable, depending on the zone they inhabit. Twilight zone fish typically have large eyes, and some focus upward to take advantage of the shadows of prey caused by the downcoming light. In the black bathypelagic zone, the eyes are usually tiny. Most whalefish, for instance, have eyes with a maximum diameter of about 2 millimeters (¹⁄₁₆ inch); there is no lens, so no image can be formed.

In this lightless environment the lateral line system, which picks up pressure waves created by swimming animals, is the most important sense organ and may be very large. In some bathypelagic fish the nasal organs are most highly developed in males, indicating that mate selection may involve female pheromones. The ear stones or otoliths of most lanternfish are large and sculptured, suggesting that sound reception may be important in this group of twilight-zone fish. However, ear stones of most bathypelagic species are small and featureless.

REPRODUCING THE SPECIES

The male and female sexes of twilight-zone fish are usually separate individuals. They breed seasonally and produce a moderate to large number of small eggs; their larvae inhabit the food-rich upper waters. In the harsher bathypelagic zone, striking modifications have occurred. All male anglerfish are dwarfs without lures. Some become parasitic on the females; after the male bites into the skin of a female, a placenta-like connection forms around the male's mouth and all nutrients are received from the female. Males and females are thus guaranteed to be together in the breeding season. The few male whalefish known are also dwarfs, less than 5 centimeters (2 inches) long, so they do not compete for resources with mature females, which may be up to 40 centimeters (16 inches) long.

A number of pelagic deep-sea fish, particularly among eel families, breed only once and die. Some bottom-dwelling fish are hermaphroditic, with both sexes present in one individual. These rare species do not have to find the opposite sex for a successful mating, just another of their own species; and self-fertilization of their own eggs may be possible.

There is still much to learn about the biology of deep-sea fish. Great advances should be made when we can keep some of these extraordinary animals alive in aquaria. ●

DIVISIONS OF THE MARINE ENVIRONMENT

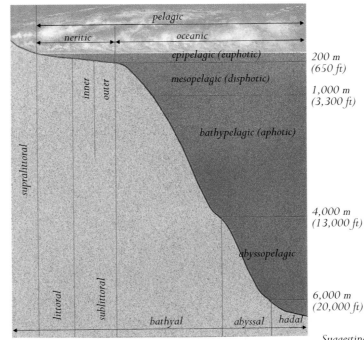

Suggesting some alien monster dreamed up by a Hollywood special-effects team rather than a real live fish, a viperfish Chauliodus displays the features characteristic of many deep-sea inhabitants: blind eyes, a relatively enormous mouth, and formidable teeth.

David Doubilet

PART THREE

OCEANS AND PEOPLE

Humans have always regarded the ocean with awe and fascination. Over the centuries, explorers have risked their lives to cross, chart, and plumb the depths of the world's oceans. Today, the oceans are an important source of food, minerals, and energy.

7 Exploring the Oceans

COLIN MARTIN

The story of oceanic exploration is one of reckless adventure and meticulous planning, honored success and tragic failure, accident and persistence, endurance and exploitation, plunder and scientific enquiry. Oceanic explorers were—and still are—motivated by forces as complex as the individuals themselves. Curiosity and commercial advantage, the thirst for knowledge and the quest for trade, patriotic pride and personal aggrandizement: all have inspired humans to cross and chart the world's oceans.

Ronald Sheridan/Ancient Art & Architecture Collection

This plate depicts Dionysus, the Greek god of wine and fertility, sailing across the sea. Dionysus is reputed to have traveled widely, perhaps as far as India, teaching people how to cultivate grapes and make wine.

Opposite. The best-documented exploratory feats are not necessarily the most epic: Polynesian seafarers, using vessels similar to these, discovered and colonized most of the tiny islands scattered across the vast southwestern Pacific.

The ancient Egyptians traveled extensively by sea to trade with other countries. The ships shown here were powered by rowers with oars, but by 3200 BC the Egyptians had invented sails.

THE FIRST EXPLORERS

Most of the explorers of the oceans are anonymous. The dates and purposes of their voyages, the vessels they sailed in, and the hazards they faced are all unknown to us. But we do know that in remote prehistory humans were capable of making long sea-crossings.

CROSSING THE OCEANS

The first Australians were aboriginal only from the perspective of the Europeans who encountered them two centuries ago: they themselves had arrived by sea at least 40,000 years earlier. Archaeologists can document the spread of early peoples from Asia along the island chain of modern Indonesia to New Guinea. There, axes and other stone artefacts have been found sealed under a layer of volcanic ash deposited about 40,000 years ago. Within 12,000 years (and perhaps considerably earlier) people had crossed the 160 kilometers (100 miles) of open water between New Britain and the Solomons.

Such voyages would have required quite sophisticated craft, capable of bearing the vagaries of wind, wave, and current in these unpredictable seas. Perhaps more significantly, their instigators would have had to break through the psychological barrier of voyaging out of sight of land, and the intellectual one of navigating without fixed landmarks.

Once these momentous barriers had been broken, the traversing and exploration of the wider ocean was just a matter of time. Certainly by 1500 BC those who had colonized the western margin of the Pacific and developed farming there had brought their root crops, livestock, and cultures deep into Oceania. By the beginning of the Christian era voyagers had reached the Marquesas. A thousand years later all the major Pacific islands lying in the great Polynesian triangle bounded by New Zealand, Easter Island, and Hawaii had been colonized by masterly seafarers—surely the most remarkable feat of sustained oceanic exploration in history.

Humans arrived in America at least 12,000 years ago via the Bering Strait. Twice over the past 38,000 years lower sea levels have created a "land bridge" across the 85 kilometer (53 mile) wide passage, and some may have crossed on those occasions. But it is likely that many came by sea. When Columbus reached America in 1492 water transport was in widespread use throughout the continent and its coastal margins from Alaska to Tierra del Fuego.

ANCIENT VOYAGERS

Maritime exploration in the Old World, radiating from the great civilizations of the Mediterranean and Mesopotamia, was mainly driven by a desire to trade. In 1750 BC Pharaoh Mentuhotep III sent an expedition down the Red Sea to Punt (modern Somalia). Two centuries later Queen Hatshepsut sent five ships there "to bring back all goodly fragrant woods . . . heaps of myrrh resin, with fresh myrrh trees, with ebony and pure ivory, with green gold of Emu, with cinnamon wood, Kheyst wood, with two kinds of incense, eye cosmetic, with apes, monkeys, dogs, and with the

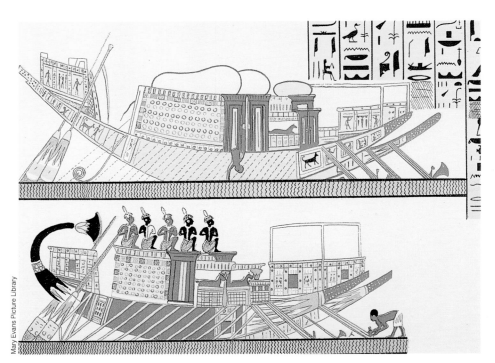

Mary Evans Picture Library

skins of the southern panther, and with natives and their children". One of her trading ships is depicted on a relief from a tomb at Kenamon, Thebes, and the accurately observed frieze of Red Sea fauna swimming below it inspires confidence in the representation of the ship.

According to the Greek historian Herodotus, Pharaoh Necho (610–594 BC) sent an expedition manned by Phoenicians to circumnavigate Africa from the Red Sea to the Mediterranean. A century later a Carthaginian expedition, attempting the voyage in the opposite direction, may have reached the Guinea coast. Another attempt was made about 470 BC by Hanno of Carthage, and although he too failed to complete the journey he does seem to have reached a point in equatorial West Africa and founded colonies there.

Northward from the Pillars of Hercules (Straits of Gibraltar) Phoenician influence centered on Cadiz, perhaps from as early as 1000 BC. By 700 BC this great port was handling tin and amber from the almost unknown lands of northern Europe. Herodotus, writing around 460 BC, speaks of these commodities as coming from "the ends of the earth"; in the same passage he mentions the "Tin Islands", which some have equated with Britain's tin-rich Cornish peninsula, though the identification is far from certain.

About 325 BC Pytheas of Marseilles sailed to Britain by way of the west coast of Spain, visiting and reporting upon the Cornish tin mines before going on to complete a circumnavigation of Britain, visiting Ireland and perhaps touching on the coast of Norway before returning home. Four hundred years later Julius Agricola, during his attempt to bring Britain's furthest extremities under Roman control, sent his fleet on a voyage of exploration around Scotland. At one stage it was accompanied by a Greek scholar, Demetrius of Tarsus, who later met the writer Plutarch at Delphi and told him of his adventures in remote Caledonian seas.

EAST AND WEST

Elsewhere in the world trade by sea between the Persian Gulf and the Indus had been established by the third millennium BC. Some two thousand years after that time extensive seafaring was also taking place along the coasts of Southeast Asia. India was the natural link between these two great spheres of maritime activity, and by the later classical period long-distance trade between west and east was well and truly flourishing. Roman coins and pottery have been found in southern India, and the spread of Graeco–Roman objects extends from East Africa to Indo-China. When the Europeans embarked upon their so-called "Age of

Discovery" in the fifteenth century, most of the world had already been discovered by seafaring travelers. It remained only for the various cultures to discover one another.

CROSSING THE ATLANTIC

The first European explorers of the Atlantic were unknown prehistoric travelers who, in bark- or skin-covered vessels, plied its coastal waters from southern Spain to Arctic Norway. In the early sixth century St Brendan, an Irish monk, undertook a remarkable expedition in a large hide-covered *curragh*, the detailed construction of which was described in a medieval account of his voyage. It seems likely that these monkish seafarers, seeking solitude in a "promised land of the saints", reached Iceland, and possibly even Newfoundland.

THE NORSEMEN

Two hundred years later the Viking expansion westward brought sturdy clinker-built ships into north Atlantic waters. The stepping-stones of Orkney, Shetland, the Faroes, and Iceland carried Norsemen to Greenland by AD 982, and within 20 years the first Europeans (so far as we know) set foot in North America when Eric the Red and his men landed somewhere in Labrador. Colonization followed, but it was short lived. Archaeologists have located the site of one Norse settlement at L'Anse aux Meadows, Newfoundland. Within a few years of its establishment, however, the site was abandoned. Perhaps the colonists were decimated by Indians; perhaps they simply could not survive so far from their parent culture.

COLUMBUS AND HIS FOLLOWERS

It fell to Iberian explorers to conquer the Atlantic, and with it the vast resources of the New World. Christopher Columbus was a Genoese of humble birth who made his momentous voyage of discovery on behalf of the Spanish crown in 1492. He had hoped to discover a westward route to Asia; instead he found the islands of the Caribbean. Three further voyages followed, the last in 1502–03, still with China as the ultimate goal.

Columbus' voyages were the inevitable outcome of the advances in science, technology, and rational thought that characterized the European Renaissance. Many of these advances, however, were rooted in the achievements of earlier civilizations. From China came the compass and gunpowder; from Arabia the skills of

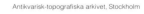

Antikvarisk-topografiska arkivet, Stockholm

Despite their extraordinary voyages, the Vikings were essentially raiders rather than explorers, roaming the seas in search of land and treasure with the full intent of taking either or both by force of arms. Weapons such as swords, shields, and helmets were therefore a standard part of any Viking expedition's equipment. This Viking helmet is of iron, the cap covered by bronze plates bearing intricate impressed designs and images of fighting men.

Opposite. *In their success as raiders and voyagers, the Vikings owed a great deal to their superb ships, which were technologically far ahead of anything else in Europe at the time. The Vikings were most probably responsible for the invention of the rudder, and they also made significant contributions to the development of the keel and the sail.*

Archiv/Photo Researchers

*"The first landing of Kolumbus",
a print by Theodore de Bry
(1528–98) showing Columbus on
the beach at Fernandez Bay in the
Bahamas. Columbus named this
region the West Indies, as he
mistakenly believed that he had
sailed as far as India.*

mathematical computation and navigational
instruments—particularly the astrolabe, by which
latitude could be gauged.

Maritime traditions that had matured for thousands
of years in the Mediterranean, refined and adapted to
Atlantic conditions by seafarers (from the Carthaginians
onward) who had sailed beyond the Pillars of Hercules,
provided a sound pedigree for the workaday Iberian
fishing boats and traders which, duly adapted, were
capable of trans-Atlantic voyaging. After Columbus the
Old World and the New were inextricably linked.

Between 1499 and 1505 as many as 11 small Spanish
fleets followed in Columbus' wake, and colonization
of the Caribbean basin began. Their outward route
exploited the summer winds that sweep southward

down the coast of Africa to the Canaries, and then
curve westward towards America. Home-bound ships
could pick up the returning winds which carried them
north of the Azores and thence to the Iberian coast.
But many did not return. The true explorers were
not the handful of individuals whose names grace the
history books but the legion of anonymous voyagers
who, from their tenuous bases on what must have
seemed the edge of the world, spread out to explore,
exploit, and colonize.

What may be the remains of one of their vessels
has been found stranded on a remote reef in the
Turks and Caicos Islands. Nautical archaeologists have
recovered fragments of the hull of a small, sleek vessel,
heavily armed with wrought-iron guns. Dating

evidence indicates that the ship went down before 1513. It probably belonged to a group of tough, freebooting Europeans intent on carving out their own slice of the New World's riches. Like modern astronauts they must have lived very close to the limits of their resources. They even carried with them the means to manufacture their own cannon-balls, underlining the one overwhelming advantage they had over the indigenous peoples of the Americas— firearms. This particular group evidently failed, but the combined efforts of their contemporaries and those who followed them brought about—for good or ill— Europe's mastery of the Atlantic.

MASTERY OF THE ATLANTIC

Other Europeans had pioneered different routes to the Americas. In 1497 John Cabot, sailing from Bristol, reached Newfoundland, thus reopening the northern route which had lain dormant since the days of Eric the Red. Amerigo Vespucci, in 1499, discovered South America, and over the following three years explored much of the coastline as far as Rio de Janeiro and possibly beyond. These discoveries were rapidly consolidated. Spain soon monopolized the Caribbean and Central America, and the mid-Atlantic routes that served them, while Portugal held sway over the eastern coastal tracts of South America.

In the north the great fishing and whaling resources off the Newfoundland coast were exploited by European adventurers, particularly the Basques, while Dutch and English explorers, vainly seeking a northwest passage to Asia, discovered and charted the closed Arctic seaways of Baffin and Hudson. In the late sixteenth century the Arctic regions of northern Europe were explored by the Dutchman Willem Barents, who was attempting to find a northern route to Asia. In 1596, Barents reached Novaja Zemla, an archipelago off the northeast coast of Russia. But his vessel was stranded in the Arctic ice, and he died the following year while trying to return to Holland in an open boat.

THE CHALLENGE OF THE PACIFIC

Many of those who followed Columbus to America still sought a passage to Asia. It was Vasco Núñez de Balboa who, in 1513, first crossed the Isthmus of Panama to reach the Pacific coast. Exploration of the western coasts of Central and South America soon followed, and in 1527 Alvaro de Saavedra crossed the Pacific to the Moluccas. But he was not the first European to navigate these waters. In 1519 Ferdinand Magellan, a Portuguese navigator with wide experience of Indian and Asian waters, set out from Seville with the intention of circumnavigating the globe and finding a western route to the Spice Islands. He sailed on behalf of Charles V King of Spain.

Mary Evans Picture Library

The Granger Collection

MAGELLAN'S VOYAGE

The voyage was fraught with difficulty. The fleet of five ships took almost six months to reach the Rio de la Plata, and the travelers were forced to winter at Port St Julian, at a latitude of 41°S. They spent much of the following year trying to find a passage through to the Pacific. One ship disappeared in the attempt and morale, already low, plummeted. Eventually the tortuous strait which now bears Magellan's name was discovered, and the little fleet broke through into the ocean beyond. Another ship was lost, and many of the men wanted to turn back. But Magellan persisted, and for three months the ships ran with the steady and gentle southeast trade winds which Magellan named "Pacific".

By now the voyagers were close to despair, with food stocks so low that they were forced to eat ox-hides, sawdust, and rats. But on March 6, 1521 they reached the

Above, right. Ferdinand Magellan (1480?–1521) led the first expedition to circumnavigate the world. Although he died before the journey was completed, the return of his ships provided the first proof to Europeans that the world was round. This engraving dates from 1695.

Above, left. A Spanish portrait of Columbus by an anonymous sixteenth-century painter. There is no known portrait of Columbus that was painted during his lifetime.

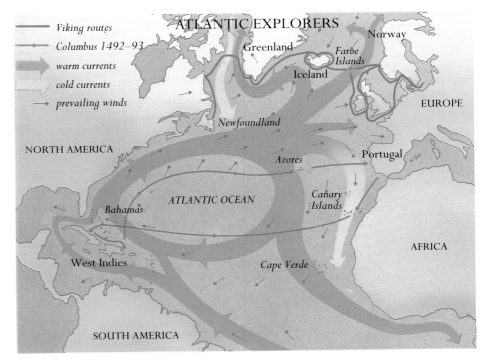

ATLANTIC EXPLORERS

— Viking routes
← Columbus 1492–93
⇒ warm currents
⇒ cold currents
→ prevailing winds

Norway
Greenland
Faroe Islands
Iceland
EUROPE
Newfoundland
NORTH AMERICA
Azores
Portugal
ATLANTIC OCEAN
Canary Islands
Bahamas
West Indies
Cape Verde
AFRICA
SOUTH AMERICA

Marianas, and the following month landed at Cebu in the Philippines. They were back in charted waters. Unhappily, Magellan was shortly thereafter killed, and the expedition was reduced to a single ship, the *Victoria*, under the command of Sebastian del Cano. When they reached the Moluccas a crewman who had been born there became, technically, the world's first circumnavigator. *Victoria* eventually reached Spain in late 1522 with only 19 remaining of the 280 voyagers who had set out. Del Cano received the honors which were rightly due to Magellan, who had achieved what Columbus had striven for—the linking of Europe with Asia via the western ocean.

SPANISH EXPLORATION

In 1565 the Spaniards pioneered the east–west Pacific route to link Asia with the Americas. This was more difficult, because the prevailing winds blow from the east. However, they discovered that north of the Philippines eastbound ships are aided by the North Pacific Drift toward a landfall in the vicinity of California. From there, the California Current runs southward to Central America. This was to become the route of the famous Manila galleons which, each year from 1574, sailed from the Philippines to Acapulco in Mexico, bringing oriental silks and porcelain in exchange for American silver, to the profit of the Spanish crown. This journey could take up to seven months, but the outward voyage was much quicker, as it exploited both the prevailing winds and the North Equatorial Current.

Other Spanish navigators such as Lope de Aguirre (1561) and Pedro de Quirós (1604–05) sought knowledge and opportunity in the Pacific, and in particular the great southern continent that was reputed to lie there. But they were not encouraged by what they found. The vastness of the ocean, and its apparent lack of continents or large islands to exploit, provided little inducement. Nor did the seafarers of other nations show much interest in the Pacific. Even Francis Drake's circumnavigation of 1577–80, during which he sailed up the west coast of America as far as Vancouver, landed in California, and then crossed the Pacific to the Philippines and East Indies, was a voyage not of exploration but of plunder. Nevertheless, he was the first to round Cape Horn, thus defining the southern end of the American continent.

Opposite. Don John of Austria's shattering defeat of the Turkish fleet at the Battle of Lepanto in 1571 marked a substantial decline in influence of the Ottoman Empire. There followed a shift to French, English, and Dutch domination, especially in trade and commerce. For a time Holland played a crucial role in exploration and cartography, as displayed in this splendid map by Pieter van den Keera, published in 1607, just as the Dutch were about to trace the coastline of Australia, the last unknown inhabitable continent.

Captain James Cook's compass. The compass, sextant, and chronometer are the three vital navigational instruments. The first two have been in use for millennia, but the last resisted the efforts of the most inventive minds until only a few years before the time of Cook's voyages. On his second voyage, Cook became the first navigator of note to leave his home port fully equipped with all three instruments. Though carefully boxed in wood, glassed over, and with its needle often suspended in oil for stability, compasses of Cook's day retained the inherent simplicity of their ancient counterparts, which were little more than a sliver of magnetized iron balanced on a pivot.

Peter Luck Productions/
Mitchell Library, Sydney

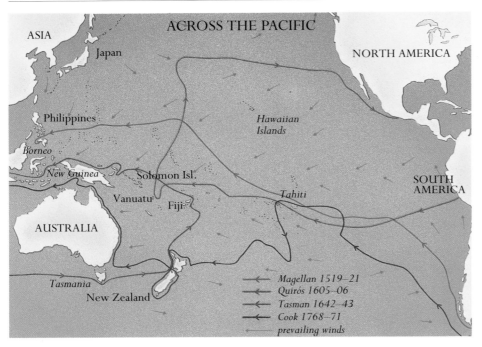

ACROSS THE PACIFIC

ASIA

Japan

NORTH AMERICA

Philippines

*Hawaiian
Islands*

Borneo

New Guinea

Solomon Isl.

SOUTH
AMERICA

Vanuatu

Fiji

Tahiti

AUSTRALIA

Tasmania

New Zealand

⟶ *Magellan 1519–21*
⟶ *Quirós 1605–06*
⟶ *Tasman 1642–43*
⟶ *Cook 1768–71*
⟶ *prevailing winds*

*Opposite. Captain James Cook's
three voyages of exploration in the
Pacific captured the imagination
of the English public, and prints
such as this one, which depicts
the* Resolution *and* Adventure
*in Papetoai Bay, Moorea, in
1777, were in demand, even
though this artist, John Cleveley,
was never in the Pacific. Cleveley's
brother, James, was a carpenter
on the* Resolution.

*A Chinese trading junk, suitable
for sailing in the open sea. This
vessel from the mid-1800s is
similar to the huge trading junks
used by the Chinese between the
ninth and fifteenth centuries for
trade on the Arabian coast and
along the east coast of Africa.*

THE EUROPEAN DISCOVERY OF AUSTRALIA

Mystery surrounds the European discovery of Australia,
the great southern continent. It is not inconceivable that
Arab or Chinese seafarers at times encountered its coasts,
whether by accident or design, while a Portuguese chart
of 1536 has been held to depict the Australian coastline
from King Sound on the Timor Sea right around the
Pacific seaboard to the southern coastline.

The Spaniards came next. Luis de Torres, who had
sailed with Quirós from Peru in 1604 but later became
separated from him in the New Hebrides (modern
Vanuatu), continued his voyage westward and probably
reached the Great Barrier Reef at about latitude 21°S.
He then headed north to round Australia's northern
tip, probably sighting Cape York but failing to recognize
it as part of the great southern continent for which he
was searching.

From the early seventeenth century Dutch East
Indiamen bound for Batavia (modern Jakarta)
regularly sighted the
Australian coast,
and some

were wrecked there. But the rugged western
coastline was of little interest to the pragmatic Dutch
merchants, who regarded it simply as a navigational
mark and a hazard to be avoided. In due course,
however, higher authority demanded fuller information
about the continent, and in 1642 Abel Tasman was sent
by the Dutch governor of Batavia to gather it. His
route took him across the Indian Ocean to Mauritius;
then, doubling back to reach Australia, he sailed too
far south and found Tasmania instead. Continuing
eastward he discovered New Zealand before returning
to Batavia via Tonga and New Guinea. In 1643–4 he
investigated the northern coast of Australia from the
Gulf of Carpentaria to the North West Cape.

COOK'S VOYAGES

The greatest European explorer of the Pacific was
Captain James Cook. In the genuine pursuit of
geographical knowledge he conducted three major
voyages between 1768 and 1779, mainly in the Pacific.
During the first he charted New Zealand, observing
the strait which separates the North and South islands.
Then, in 1770, he surveyed the entire east coast of
Australia. His second voyage aimed to establish
whether—as some believed—another great continent
lay in the south Pacific. In the course of proving that it
did not exist he visited and charted the exact positions
of Easter Island, the Marquesas, Tonga, Tahiti, and
Vanuatu, and discovered New Caledonia, Norfolk
Island, and the Isle of Pines.

Cook's third voyage resulted in the discovery of
Hawaii, and the exploration of the northern coastline
of America to the Bering Strait and beyond. He then
sailed southward along the coast of Asia before
returning to Hawaii where he met the same fate as
Magellan in the Philippines—he was killed by natives.

EXPLORERS OF THE INDIAN OCEAN

The coastal fringes of the Arabian Sea, from the Horn
of Africa to the southern tip of India, were well known
to the ancient world. But the hegemony enjoyed by
classical seafarers passed to the Arabs with the fall of
the Roman empire, and with the rise of Islam in the
seventh century Muslim merchants rode the monsoon
winds and brought their *dhows* to India, Malaya,
Indonesia, China, and East Africa. Such voyages were
the inspiration for the legendary journeyings of Sinbad
the Sailor, whose stories date back to the ninth century.

CHINESE EXPEDITIONS

Although coastal watercraft in Southeast Asia have
been used from the remote past, oceanic voyaging was
a much more recent phenomenon. By the ninth
century Chinese navigators possessed information
about the Gulf of Aden and the Somali coast, and 200
years later they knew of Zanzibar and Madagascar.

Marco Polo, the Venetian traveler, described the Chinese ocean-going junks he saw in the 1290s. They had four masts, and some also had two auxiliary ones which could be raised and lowered as required. There may have been crews of up to 300 and the largest ships could carry 6,000 baskets of pepper.

Although Chinese introspection and mistrust of foreigners discouraged exploratory voyages, a remarkable series of maritime expeditions took place in the fifteenth century. They were conducted by Cheng Ho (also known as the Three-Jewel Eunuch), admiral of the Ming emperor Yung Lo, who between 1405 and 1433 led seven expeditions to visit 37 countries on the coasts of Indo-China, the Indian Ocean, the Persian Gulf, Red Sea, and East Africa. Among the more exotic acquisitions he brought back to the emperor was a fully

grown African giraffe. But Yung Lo's successors closed the exploratory doors opened by Cheng Ho, and the road to the east through the Indian Ocean was left open to other seafarers.

PORTUGUESE EXPLORATION AND COLONIZATION
Perhaps more than anything else it was the existence of the Moslem empire that encouraged Europeans to bypass it by opening maritime routes to the east. The main inspiration came from Portugal, where Prince Henry ("The Navigator") had set up his celebrated school of navigation in 1416. By the middle of the century mariners like Alvise da Cadamosto and Diego Gomez were probing the coast of West Africa, and by 1487 Bartholomew Diaz had reached the Cape of Good Hope.

Ten years later Vasco da Gama set out from Lisbon on his epic voyage to India, touching Cape Verde before

The most common type of sailing vessel at the time of Prince Henry the Navigator (1394–1460) of Portugal was the carrack, shown in this contemporary painting. These vessels plied the trade routes of the Mediterranean but were built for capacity and profit, not versatility. One of Prince Henry's great contributions to ocean exploration was the development of an alternative vessel, the caravel, which was smaller, more maneuverable and seaworthy, and better suited for lengthy voyages.

heading west to a point some 960 kilometers (600 miles) from the South American coastline to avoid the currents in the Gulf of Guinea which had deterred navigators from Carthaginian times. He then doubled east to make a landfall just north of the Cape, from which he proceeded via Mossel Bay, Natal, and Mombassa to Malindi. There he took aboard a local pilot to convey him on the long-established Arab route to Calicut. His return trip, sadly, was one of plunder rather than exploration. The way was now clear for the Portuguese and their successors to discover and exploit the resources of Africa, India, and the Far East.

This colonial process accounts for practically all the European seafaring, and hence exploration, that took place in the Indian Ocean and beyond over the following three centuries. First came the lumbering Portuguese *náos* of the *Carreira da India*, which sustained, by their wearisome and

A portrait of Prince Henry the Navigator, who in 1416 set up his renowned fortress at Sagres, on Portugal's Algarve Peninsula. The fortress was a school and center for cartography, navigation, and shipbuilding.

Mary Evans Picture Library

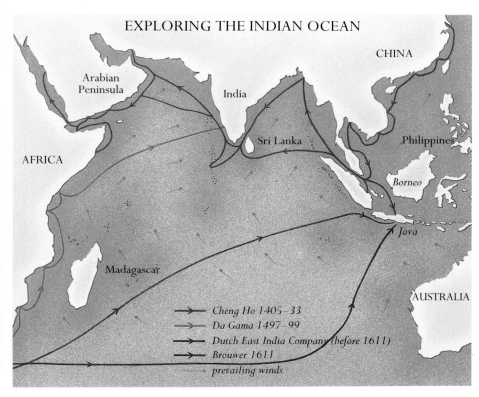

EXPLORING THE INDIAN OCEAN

CHINA

Arabian Peninsula

India

AFRICA

Sri Lanka

Philippines

Borneo

Java

Madagascar

AUSTRALIA

→ Cheng Ho 1405–33
→ Da Gama 1497–99
→ Dutch East India Company (before 1611)
→ Brouwer 1611
→ prevailing winds

dangerous six- to eight-month voyages, the farflung trading stations along the East African coast, at Goa, Calicut, and Cochin on the Indian subcontinent, and thence via Colombo and Malacca to Macao on the Chinese mainland. By the end of the sixteenth century this vast and cumbersome trading empire, now united with the American and Pacific conquests of Spain, was at the zenith of its power and prosperity.

DUTCH AND ENGLISH EXPLORERS

But there were rivals. Dutch and English seafarers, inspired by new religious ideology as well as by commercial aspirations to break the Spanish–Portuguese monopolies, and encouraged by maritime successes in their own waters, were eager to develop and control trade with the Far East. The tools with which they sought to achieve this were the great incorporated East India companies, founded in 1600 (English) and 1602 (Dutch). Their sleek ships, built in the tradition that produced the revolutionary ocean-going warships that had resisted the Spanish Armada in 1588, proved ideal for the run to India and beyond.

Dutch ingenuity in navigating the Indian Ocean led to the European discovery of Australia. In 1610 Henrik Brouwer pointed out that the usual route taken to Java was unsatisfactory, since it involved crossing in a tropical latitude, where the wind, if not contrary, was often absent and where the heat caused pitch to melt, cargoes to rot, and men to fall sick. Far better, he argued, to follow a temperate latitude from the Cape, heading north only when the longitude of Java—about 105°E—had been reached. This point is only 800 kilometers (500 miles) from the western coast of Australia.

Brouwer himself pioneered this route with conspicuous success, and thereafter it was routinely taken by Dutch East India Company vessels. It was only a matter of time before one of them discovered the Australian continent. The honor fell to Dirk Hartog of the *Eendracht*, who in 1616 reached the island off Western Australia which now bears his name. Others were less fortunate, and were wrecked. One, the *Batavia*, drove into the islands of the Houtman Abrolhos in 1629, and its remains are now on display in the Western Australian Maritime Museum.

Patrick Baker/WA Museum

POLAR EXPLORERS

The last great unexplored tracts of the world's oceans lay in the polar regions. As long as the motives for discovery were colonization or exploitation, these barren and climatically hostile areas held little appeal. But the sixteenth-century seafarers who penetrated the northern polar region in search of a northwest passage were soon followed by fishermen and whalers who seasonally visited the fishing grounds off Newfoundland and the whaling stations of Labrador and Spitzbergen. Throughout much of the seventeenth century these waters were the scene of fierce rivalry between the fishing nations, particularly the English and the Dutch, and discoveries were often jealously guarded as commercial secrets.

THE ARCTIC

By the eighteenth century geographical enquiry—though not always untouched by political motives—was more frequently conducted for its own sake, and the polar regions began to be explored. In 1728 the Dane Vitus Bering, on behalf of the Russian crown, discovered that the continents of America and Asia were separated by the strait that now bears his name. A few years later the Russian navigator S.I. Chelyuskin explored much of Arctic Russia's northern coastline. In 1806 William Scoresby, after traversing the Arctic east coast of Greenland, reached a latitude of 81°30'N in the vicinity of Spitzbergen. Twelve years later Sir John Ross sailed up the west coast of Greenland into Baffin Bay, and on a later expedition (1829 to 1833) he explored the islands and seaways of the Canadian Arctic.

These hazardous journeys sometimes ended in disaster. In 1846 Sir John Franklin, searching in the same area for the elusive northwest passage, became trapped in the ice. His party's fate was later revealed by rescue attempts and, subsequently, by archaeologists. A search party aboard the barque *Breadalbane* became icebound in 1853 and, although its members survived, the ship was lost. At length, in 1859, the remains of Franklin's party were discovered on King William Island, together with a diary recording their terrible ordeal. More than a century later, in 1984, archaeologists discovered some of Franklin's men buried in the permafrost, their flesh and features preserved and their clothing intact. About the same time the wreck of the rescue ship *Breadalbane* was found and photographed, 100 meters (330 feet) below the Arctic ice.

A northwest passage was finally achieved by the Norwegian Roald Amundsen, who in 1903 sailed from the Atlantic via the McClintock Strait to emerge through the Bering Strait into the Pacific.

Jeremy Green/WA Museum

TO THE NORTH POLE

The final prize of Arctic exploration was the pole itself. Inspired by Cook's successes in the Pacific, Britain's Royal Society commissioned two ships in 1773 to make an attempt on the North Pole. But the expedition, led by C.J. Phipps, failed to find a passage north of Spitzbergen, and narrowly escaped disaster. It was on this voyage that the 14-year-old Horatio Nelson had his celebrated encounter with a polar bear. In 1827 another attempt was made by William Parry, an outstanding seaman and explorer of Cook's caliber, who by sledging boats over the drift ice that lay between stretches of open water reached a latitude of 82°45'N.

Following an attempt by the Norwegians Fridtjof Nansen and Frederic Hjalmar Johansen in 1895, during which they came within four degrees of their goal, the pole was finally reached in 1909 by the American Robert Peary, accompanied by Matthew Henson and four Eskimos. Recently this claim has been challenged, but a mathematical resolution of the shadows cast in Peary's photographs has demonstrated that he came as close to the true pole as his instruments could determine.

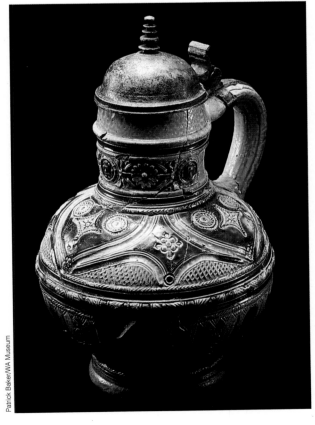

Patrick Baker/WA Museum

Above. The development of maritime archaeology in the past 50 years has told historians much about trade and exploration of the seas, and more specifically, has revealed details about the vessels that undertook these journeys and the cargo they carried. Here, a diver uncovers the hull timbers of the Batavia, *sunk in 1629.*

Opposite and left. The Dutch East India Company's operations in the East Indies reached a peak in the late seventeenth century. The large number of Dutch wrecks found on these trade routes is a reflection not only of the Dutch desire to dominate trade in the East but also it is a testament to how dangerous long ocean voyages were at the time. The wreck of the Batavia *was found on the Houtman Abrolhos Islands, off the Western Australian coast in 1963, and many artefacts have been retrieved from it, among them the ivory carving (opposite), which was probably a cutlery handle, and the Westerwald stoneware jug (left).*

Sir Robert Falcon Scott (1868–1912) and his team at the South Pole, January 18, 1912. On the return journey all five men died from hunger and cold.

Opposite. Establishing a northwest passage proved an elusive goal, and several explorers came to grief while trying to find it. Sir Allen Young commanded the Pandora *in an attempt in 1875, but his ship became ice-bound and had to be abandoned. This contemporary print shows the* Pandora *trapped in Melville Bay in 1876.*

THE ANTARCTIC

During his second voyage, in 1773, James Cook became the first person to cross the Antarctic Circle. In trying to determine whether there was a continent in the far south he was turned back by pack ice when he was no more than 160 kilometers (100 miles) from the coast of Queen Maud Land. Later that year, from the Pacific side, he reached a latitude of 71°S. Almost half a century later, between 1819 and 1821, the Russian navigator Fabian von Bellingshausen circumnavigated Antarctica, and discovered Alexander I Island. On his return journey von Bellingshausen encountered seal hunters who had just discovered the tip of the Graham Land peninsula.

Exploration along the fringes of the continent followed: Weddell in 1823, Biscoe in 1831, and Dumont d'Urville in 1840. But the major investigation was that conducted by John Clark Ross in a series of voyages between 1840 and 1843 with the ships *Erebus* and *Terror*. Having sailed from Tasmania, they penetrated the great sea (now the Ross Sea), largely free of ice, which pushes into the continent. Ashore they discovered a mountainous coast and two active volcanoes, which they named after the ships. A second season was devoted to further survey in this region, and a third was spent exploring the fringes of the Weddell Sea.

The Ross Sea provided explorers with access to the Antarctic continent at a latitude of 78°S—a distance of less than 1,600 kilometers (1,000 miles) from the pole itself. A tragic race to be the first to reach the South Pole began in 1910 between a Norwegian team under Roald Amundsen, who planned to negotiate the final leg with the aid of dog-sleds, and a British expedition led by Captain Robert Scott, who proposed that the sleds would be hauled by the expeditioners themselves. Amundsen and his Norwegian team reached the pole on December 14, 1911, and returned safely. A month later the British team got there too, but Scott, along with all of his companions, perished during the return journey. ■

8 EXPLORING THE DEPTHS

SYLVIA A. EARLE

Exploring the depths has an appeal that in many ways compares with the lure of the skies above. Throughout history, many have yearned to swim and dive like dolphins and whales. There are scattered accounts over the past two thousand years of diving for food, for recovering ships and goods lost at sea, for carrying out military strategies, and for the sheer pleasure of exploring.

Opposite. Through storms, misadventure, poor navigation, or maritime battles, the ocean floor has become a graveyard for uncountable sunken ships, most long since lost and forgotten. This wreck off the coast of New Ireland is that of a Taiwanese fishing vessel, caught poaching and sunk by the Papua New Guinea Navy. Many such wrecks become artificial reefs, home to a multitude of fish and encrusting organisms.

Apart from its importance in a wide range of purely technological applications, the invention of scuba gear has also promoted vigorous development in the marine biological sciences, allowing researchers to study directly the structure and behavior of marine animals such as this potato cod.

THE BEGINNINGS OF OCEANOGRAPHY

Alexander the Great is said to have been so taken with the notion of penetrating the depths that in 433 BC he had a diving bell constructed with a glass viewing window. The breath-holding exploits of the Japanese and Korean *ama* divers, mostly women, have been depicted in artistic renderings for more than 1,000 years, and continue even now with only modest concessions to modern technology. During the past several centuries, many attempts have been made to develop equipment to improve access to the ocean depths. Various diving bells were designed and used during the 1600s, and in 1715 successful salvage dives to 20 meters (65 feet) were made in a cylindrical wooden barrel equipped with a small viewing port.

The use of compressed air for diving began in the late 1700s, at the same time as the development of metal helmets and flexible suits. This technique continued during the 1800s, particularly following the appearance of a diving suit and helmet designed by Augustus Siebe in 1819. In the mid-nineteenth century, Jules Verne wrote convincingly of self-contained diving systems in his classic work, *Twenty Thousand Leagues Under the Sea*. As is often the way, science fiction anticipated science fact, in this case by many decades.

Jacques Cousteau and Emile Gagnan perfected the first practical, fully automatic compressed air "aqua-lung" in 1943, and self-contained underwater breathing apparatus—"scuba"—soon became an accepted technique for diving. Leonardo da Vinci and Benjamin Franklin both conceived designs for swim flippers many years before an ideal material—rubber—became available in the 1930s.

Most of the equipment that makes modern diving possible has been developed in recent decades. Millions of sport divers now don masks, fins, snorkels, scuba, and an impressive array of accessories for excursions to 50 meters (165 feet), sometimes more. Scientific, commercial, and military dives are possible to depths over 500 meters (1,650 feet), using exotic mixes of gases, special suits, and chambers for underwater living and decompression.

A few experimental dives have exceeded 600 meters (1,970 feet), but the ability to probe significantly deeper using techniques that expose humans directly to the dark, cold, high-pressure environment of the deep sea is proving elusive. Other approaches involving protective suits and submersibles, and a host of remotely operated underwater vehicles are more promising as a means of gaining access to the full range of the ocean's depth, even the deepest sea.

THE VOYAGE OF THE *CHALLENGER*

Considering the relatively recent development of practical diving equipment, it is no wonder that early oceanographic expeditions concentrated on the sea surface, and used equipment that could be lowered from the deck of a ship to gather information about the depths below. Oceanography—the systematic scientific exploration of the sea—began in earnest with the departure of *HMS Challenger* from the shores of England in 1872. A four-year expedition to the major oceans of the world followed, an epic journey launched to answer such fundamental questions as: How deep is the ocean? Is there life in the deepest sea? Where do oceanic currents go? How salty is the sea?

Challenger scientists returned with many answers; but many more questions were provoked by the discoveries they made. Nodules of manganese were dredged from the deep-sea floor. Why are they so

Quinn/Australian Picture Library

Two modern divers use facsimiles of sixteenth- and seventeenth-century equipment in a demonstration of early salvage techniques. In both cases, the diver's activities take place outside the bell, the diver returning to its shelter to take another breath. The seventeenth-century Sturmius bell (right) allowed a diver to remain submerged for up to 20 minutes before its reservoir of air was depleted, but Tartaglia's earlier device (above) permitted dives that lasted for no more than a few minutes.

A DEVELOPING SCIENCE

A comprehensive study of the south Atlantic during the 1925–27 voyages of the oceanographic vessel *Meteor* set a new standard for oceanic research. Its scientists were among the first to use an electronic echosounder to measure ocean depths, a technique that provided revolutionary insight into the configuration of the planet's surface. More than 70,000 echosoundings clearly showed the rugged nature of the sea floor, and subsequent investigation by many expeditions revealed a startling structure of mid-oceanic mountain ranges, extending like an immense backbone down the Atlantic, Indian, and Pacific oceans.

Challenger scientists were unaware that one of the ocean's dominant features had escaped their notice. Upon reflection, however, the wonder is that scientists a century ago learned as much as they did, considering the technology available. Even now, oceanographers can be likened to detectives, working with fragments of information to put together the pieces of an enormous and ever-expanding puzzle.

Consider the problems. If standard oceanographic techniques were applied to exploration of cities or forests or mountain ranges, what would we know about these places? Suppose a net were dragged blindly through the streets of a town and the items randomly captured emptied onto the deck of an aerial equivalent of a research vessel. The resultant jumble of shrubbery and cement, with perhaps an automobile or confused pedestrian, would give some insight about what occurs below, but real understanding would be as elusive as real understanding is today of deep-sea communities.

New cameras, submersibles, robots, and other technology are greatly increasing our knowledge of the oceans. The *Challenger* expedition learned more about the sea than had been discovered during all preceding history. In the past half-century, it is safe to say that, once again, more has been learned than during all preceding human history. With new voyages and new technology, it is likely that the greatest era of oceanic exploration has just begun.

GRAND CANYONS OF THE DEEP

The greatest depths of the sea are in the steep-sided, narrow, deep-sea trenches that form a nearly continuous border not far offshore from the coastal ranges of the Asian and American continents. The upper edge of most of these grand canyons of the deep begins at about 6,000 meters (20,000 feet)—about half the maximum depth of the sea. Access to these unique areas has been and still is extremely difficult. Only once has a descent been made to the bottom of the deepest place, the Challenger Deep in the Mariana Trench near the Philippines.

abundant? How were they formed? Answers are not yet forthcoming. Thousands of species of plants and animals new to science were discovered, and thousands of measurements were made of the ocean's temperature, salinity, and chemistry. Samples of rocks and mud and sediment were taken from the Atlantic, Pacific, and Indian oceans. They provided many clues but few real answers about the basic physical character of the ocean.

In the more than 50 volumes that reported on the extensive findings of the *Challenger* expedition, an underlying message emerged concerning both the great wealth of information and the magnitude of unknowns remaining about the sea. Numerous expeditions from many nations followed, but not until the early twentieth century did a general understanding of the oceans begin to take form.

Using the bathyscaphe *Trieste*, US Navy lieutenant Don Walsh and Swiss engineer Jacques Piccard made their historic journey in January 1960, resolving both the question of whether or not such a descent could be made (it could) and whether or not life occurs in the deepest sea (it does). Looking back at the two explorers, almost 11,000 meters (36,000 feet) down, was a flounder-like fish, at home in a realm where pressure is 110,240 kilopascals (16,000 pounds per square inch).

Using various acoustic sounding techniques, the general configuration of the deep-sea trenches has been charted, but taking samples and even photographs in these great gashes in the sea floor presents enormous challenges. Oceanographer Willard Bascomb is among those who have spent considerable time trying—and sometimes succeeding. His recent book, *Crest of the Wave*, describes his exploration of the Tonga Trench, a formation about 2,400 kilometers (1,500 miles) long, 24–48 kilometers (15–30 miles) wide, and almost 11,000 meters (36,000 feet) deep.

To sample this magnificent crack, eloquently described as "a mile deeper than Mount Everest is high" or, viewed another way, "as deep as seven Grand Canyons but with much steeper sides", Bascomb and his colleagues lowered a gravity corer into the depths. The corer is an instrument about 2 meters (6½ feet) long, weighted with lead, that is used to obtain a cylindrical plug of whatever is on the bottom. When it was retrieved, the sampler was empty, but embedded in the lead weight was a chip of basalt. This represented a slight but important clue from which deep-sea oceanographic detectives could extrapolate the nature of one of the planet's great deep-sea formations.

A QUEST FOR THE FUTURE

Since Walsh and Piccard's visit to the bottom of the Mariana Trench, little attention has been given to direct access. Most deep-sea trenches start at a depth where even the deepest modern submersibles must stop. It seems curious that no present submersible can directly travel to the deepest sea, thus facilitating exploration of the deep-sea trenches. While constituting only approximately 3 percent of the total ocean, the depths below 6,000 meters (20,000 feet) represent unique areas with high scientific value. They are also recurrently posed as possible sites for disposal of toxic wastes. Some argue that exploration must precede such use, to avoid possible disruption of systems that may have high value beyond waste disposal.

Three percent may sound very small, but this represents approximately 10 million square kilometers (more than 3½ million square miles), an area about the size of Australia or the United States or Canada. It may be dark and cold and remote with a unique high-

Al Giddings/Ocean Images

pressure environment where strange creatures dwell—but it is a special part of the planet currently inaccessible for scientific exploration.

Interest in exploration of the deep-sea trenches is growing. In Japan, development is underway for construction of a large tethered remotely operated underwater vehicle that will be equipped with cameras and various sampling tools to operate in depths to 10,000 meters (33,000 feet). In the United States, efforts are focused on small manned submersibles planned as part of a project known as "Ocean Everest". In an era when traveling 11,000 meters (36,000 feet) high in aircraft is routine, and spacecraft can go to the moon and beyond, it seems likely that direct access to the full ocean depths, including the deep-sea trenches, will soon be possible.

Typical of most other submersibles, the Johnson Sea-Link *is specifically designed to move slowly and observe closely. Accessories can be bolted on wherever they can be used most effectively, and do not have to conform to an overall efficient shape. The vehicle bristles with lights; measuring instruments; still, movie, and video cameras; sampling devices; and manipulators of various kinds, including two articulated arms.*

Opposite. *Mini-ROVER hovers above a giant spider crab* Macrocheira kaempferi. *Some specimens have been found that measure more than 3 meters (10 feet) from claw-tip to claw-tip, although it may take them up to 50 years to reach such a size. Known as "shinin gani", or dead man's crab to Japanese fishermen, it is confined to deep waters off the coast of Japan, where its usual diet of mollusks, crustaceans, and fish is sometimes supplemented by the corpses of drowned humans.*

SUBMERSIBLES: SPACECRAFT OF THE SEA

Modern submersibles may be likened to "inner spacecraft", the undersea equivalent of advanced systems that transport astronauts skyward, beyond the atmosphere familiar to humankind. Similarly, it is necessary to have protective vehicles for those who wish to travel to the hostile atmosphere of the deep sea, or to use remotely operated vehicles, robots, and telepresence—the undersea analogs of unmanned space probes and satellites.

One of the pioneers of deep-sea exploration, Dr. William Beebe, was so impressed with the parallels between the sea and space that he wrote in his historic work, *Half Mile Down*, that the only place comparable to the deep sea "must surely be naked space itself, out far beyond atmosphere, between the stars . . ." Beebe likened the dazzling array of deep-sea creatures dwelling in the vast darkness, glistening in the lights of his undersea vehicle, to "the shining planets, comets, suns, and stars".

Without submersibles, this magnificent realm would be known only indirectly, and access to the sea would be limited to the depths divers can reach. Scuba diving is effective to depths of about 50 meters (165 feet), and other diving methods can be applied for limited access to more than 500 meters (1,650 feet). But it is well to keep in mind that the average depth of the sea is at about the depth where the sunken passenger liner *Titanic* is resting on the sea floor—more than 4,000 meters (13,200 feet) beneath the ocean's surface.

Fortunately, in the half-century since Beebe's early exploration, numerous undersea vehicles have been developed and used to help explore the ocean depths. Two submersibles have visited the site of the *Titanic*: the research vessel *Alvin*, operated by Woods Hole Oceanographic Institution, and the French vehicle, *Nautile*. Although more than a quarter of a century has passed since *Alvin* was launched, the system has been renovated many times, and has maintained the most modern instrumentation. During exploration of the wreck of the *Titanic*, a small remotely operated vehicle, *Jason Jr*, was launched from *Alvin* and, operated with remote controls from within the submersible, was able to probe inside the shipwreck in places too small and too dangerous for access by *Alvin* directly.

The combination of a manned submersible working effectively with a small remotely operated vehicle highlights a point of view shared by many who work in the ocean. The issue is not whether manned systems are better than remotely operated or robotic ones; rather, that the best tool for a particular job should be chosen, and sometimes both will work together.

Emory Kristof, Michael Cole, and Keith Moorehead/National Geographic Society

HOW DEEP CAN THEY GO?

In recent years the ocean floor has been mapped by surface sonar and similar remote techniques, but only a fraction of the ocean depths has been explored in any direct physical sense. Until very recently, most technological effort has been focused on the design and fabrication of equipment that can safely and reliably withstand the enormous pressures encountered at such depths. To date, about 600 meters (1,950 feet) marks the limit of direct human exposure to the ocean environment, and greater depths can be reached only by submersible vehicles.

1 The Conshelf (Continental Shelf Station) series of experimental installations were designed to explore the limitations of humans living and working at high pressures for extended periods. In 1965, in Conshelf Three, six men remained for 22 days at 100 meters (330 feet) in the Mediterranean.

2 Fast-moving open-ocean fish like tuna seldom descend further than about 200 meters (650 feet).

3 Frenchman Jacques Cousteau's pioneering Diving Saucer was taken to a depth of 412 meters (1,350 feet) in 1959.

4 The JIM suit is theoretically capable of descending to 610 meters (2,000 feet), but 439 meters (1,440 feet) is the maximum depth so far reached in any operational dive.

5 The Johnson Sea-Link can operate at depths up to 600 meters (1,950 feet), and is equipped to release and retrieve diving personnel.

6 The US Navy's experimental NR-1 nuclear submarine can descend to about 700 meters (2,300 feet) and move on the bottom on wheels.

7 Giant squid spend much of their lives at depths of 1,000 meters (almost 3,300 feet) or more.

8 The US Navy's experimental submarine AGSS 555 (Dolphin) reached 1,525 meters (5,000 feet) in 1988. It is by far the largest vessel illustrated here.

9 In 1934, Dr. William Beebe and designer Otis Barton took their bathysphere to a depth of 923 meters (3,028 feet) near Bermuda, a world record at the time.

10 Specially designed to rescue trapped personnel from disabled submarines, the US Navy's DSRV-1 (Deep Sea Rescue Vessel) was successfully tested at 1,525 meters (5,000 feet) in 1988.

11 In 1975 the French submersible Cyana carried a three-person crew to around 3,000 meters (9,800 feet).

12 Operated by the US Navy, DSV-3 (Turtle) has an operational depth limit of 3,050 meters (10,000 feet).

13 Still operational after more than 25 years, though refitted several times, Alvin can dive to a depth of about 4,000 meters (13,000 feet).

14 The US Navy's DSV-4 (Sea Cliff) was tested at 6,100 meters (20,000 feet) in 1988.

15 Operated by the Scripps Institution of Oceanography, Deep Tow is a deep-ocean sampling and instrument package controlled by and tethered to a surface research vessel by a long cable. It has been successfully used at depths up to 6,100 meters (20,000 feet).

16 In 1960 US Navy lieutenant Don Walsh and Swiss engineer Jacques Piccard took the Trieste to the deepest point in the world's oceans, 10,860 meters (35,630 feet) in the Mariana Trench near the Philippines, safely returning with two items of information: the ultimate descent is achievable, and life exists even in the deepest parts of the ocean.

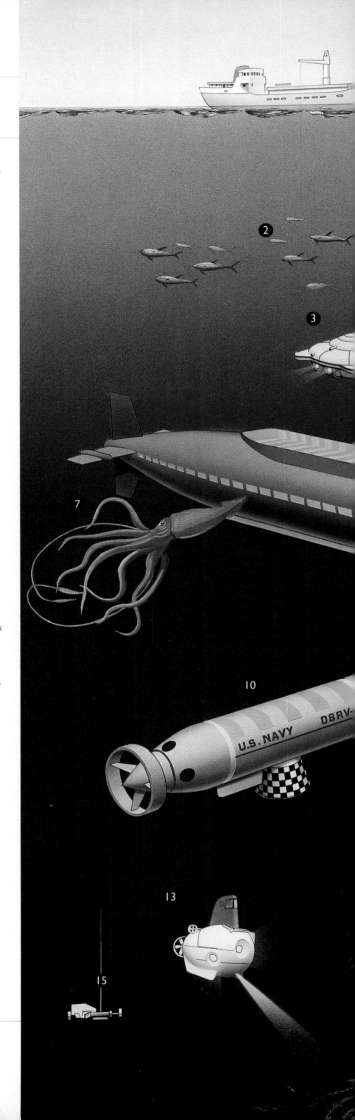

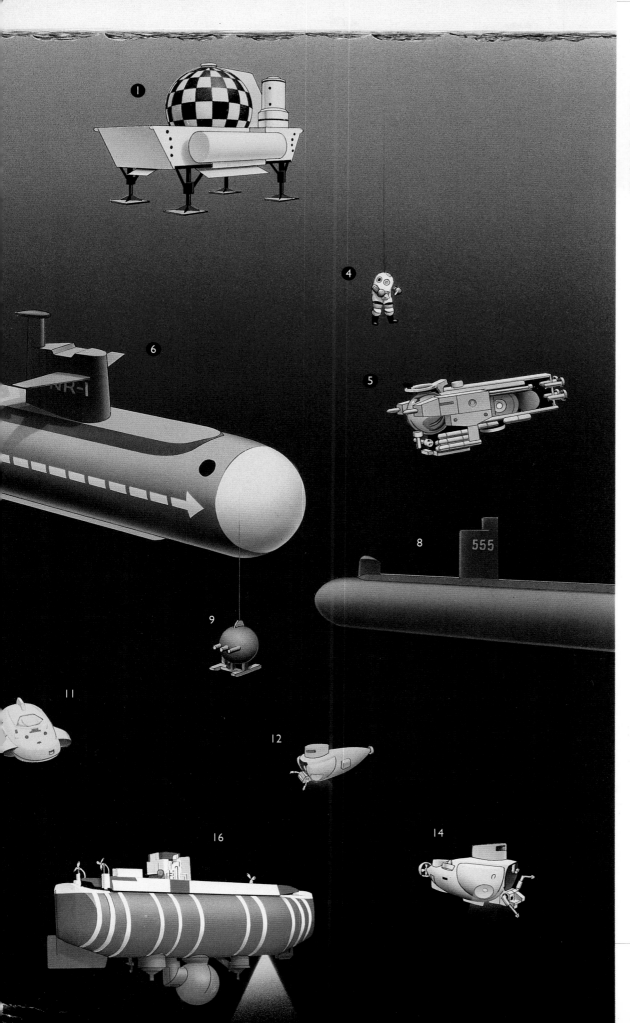

William Smithey/Planet Earth Pictures

Operated by the Woods Hole Oceanographic Institution in Massachusetts and renovated many times, the submersible Alvin has been the workhorse of deep-ocean exploration since its first dive in 1964.

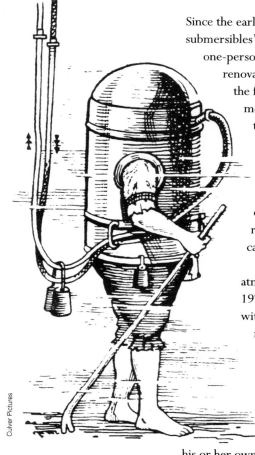

Early diving gear was little more than a metal hood enclosing a bubble of air pumped from the surface through a long flexible hose. The fact that the air pressure must therefore match the pressure of the surrounding water severely restricted the depths that could be reached, as well as rendering the diver vulnerable to "the bends" on ascending. Such "multiple atmosphere" technology is still used today, now that effective counters to both difficulties have been devised.

Opposite. The JIM suit is currently the extreme development of "single atmosphere" technology. Usually tethered to a surface vessel, its self-contained life-support systems maintain internal pressure at sea level (one atmosphere) regardless of depth. In 1976 a diver wearing one of these suits descended to 439 meters (1,440 feet) to recover a submarine cable off the coast of Spain.

PERSONAL SUBMERSIBLES

Since the early 1970s the rapid evolution of "personal submersibles" has greatly altered ocean access. These one-person, one-atmosphere systems began with a renovated version of a system called *JIM* named for the first person to wear one. Like larger submersibles, *JIM* has self-contained life support that is maintained at surface pressure—one atmosphere. It is made of articulated metal sections joined by special seals. Normally operated on a tether to a surface ship or platform, one of the 15 modern versions of *JIM* was used in 1979 for exploratory research dives in Hawaii without the use of a cable to the surface.

Other tethered one-person, one-atmosphere systems were developed in the 1970s and 1980s, but one that operates without a tether has become especially renowned. Named *Deep Rover*, the vehicle is basically a clear acrylic sphere that looks much like the front end of the *Johnson Sea-Link* submersibles. For precise work, two large sensory manipulators are used by the operator as almost instinctively controlled extensions of his or her own arms and hands. It is a splendid example of the interaction between human and machine, and is so "user friendly" that piloting it has been likened to driving a golf cart. This characteristic, coupled with its small size and lack of a tether, makes it possible to use *Deep Rover* for scientific operations in remote locations inaccessible to large submersibles or even tethered remotely operated vehicles that require complex cable management.

Late in the 1980s, increasing interest in oceanic research resulted in the development of several new systems that can dive to 6,000 meters (20,000 feet)—and thus to approximately one-half the ocean's depth, but within reach of about 97 percent of its area. In 1987, the Soviet Union launched two three-person submersibles, *Mir I* and *Mir II*, built in Finland and operated worldwide from a large research vessel. Japan now operates the three-person *Shinkai* 6500, the deepest diving submersible currently available. The French submersible, *Nautile*, and the US Navy's recently renovated *Sea Cliff* complement the more modern members of the global fleet of five deep-diving manned systems.

UNMANNED SUBMERSIBLES

Unmanned submersibles—remotely operated vehicles and autonomous underwater vehicles—have been edging to the forefront of oceanic exploration and research. Exotic communities of life and associated hydrothermal vents in the deep sea near the Galapagos Islands were first glimpsed through the television "eyes" of an underwater robot, a towed array of instruments operated from a surface ship.

Later, *Alvin* transported scientists directly to the site for first-hand observations and documentation. *Titanic* was found by a towed system equipped with special cameras and lights and *Alvin* later took observers to see at close range what the remotely operated system had discovered.

The first submersible to dive deeper—and much longer—than divers with air tanks in the Antarctic was not a manned submersible, but one of the nearly 200 portable *Phantom* remotely operated vehicles currently being operated globally in depths as great as 900 meters (3,000 feet). Tetherless systems equipped with computer controls are being touted as the cutting edge of submersible technology, and several variations are being developed in the United States, Britain, Canada, Japan, and elsewhere.

THE WAY FORWARD

New materials, new power sources, and new incentives are the driving forces for innovations in submersible technology that should revolutionize access to the depths. Part of the revolution will be attributable to the rapid acceptance and use of remotely operated vehicles and underwater telepresence but even for manned systems, the changes are expected to be dramatic.

Until now, most manned submersibles have operated on the same principles that govern the movement of balloons and dirigibles—a variable ballasting system. The submersibles descend with weights, either water that is later displaced with air for buoyancy, or lead or iron weights that are left behind to enable the vehicle to become lighter than the surrounding water, and thus able to ascend. The *Mir* submersibles, *Alvin*, *Deep Rover*, and many more operate this way.

The advent of new materials, new batteries, and new designs has led to another approach, one that can be likened to a small, fixed-wing aircraft. Two small, portable, one-person, battery-powered systems called *Deep Flight* have been constructed of "modern" materials—acrylic, glass fiber, composites, and epoxy—and without ballast tanks. Designed to be literally flown through the water, *Deep Flight* is expected to be used for unique scientific applications such as swimming beside whales and fast-moving fish, and conducting surveys in remote areas not readily accessible to larger systems. The *Deep Flight* design should also serve as a prototype for deeper, faster vehicles that will use new power sources and modern non-metallic materials for systems that will be able to travel freely throughout the ocean.

What are the limits of submersibles in the sea? The potential of existing technology has not yet been fully realized, and with the addition of new through-water communications, new power sources, new strong, light materials, and a powerful incentive to understand how the oceans work, the developments should proceed rapidly toward a "space age" in the sea. ■

9 FOOD AND ENERGY FROM THE SEA

CHRISTIAN D. GARLAND, JAMES C. KELLEY, AND G.L. KESTEVEN

*A*lthough the oceans cover about 70 percent of the Earth's surface, the marine contribution to human food supplies is much smaller than that from land. The world harvest of agricultural products is perhaps 30 to 40 times as great as that of fishery products. Nevertheless, the oceans are an important source of food—obtained either by harvesting from the wild or by farming. Moreover, the sea provides significant amounts of oil, minerals, and renewable energy, the extraction of which is both a technological challenge and, perhaps, a pointer to our future.

HARVESTING THE SEA

What is obtained from a resource, as distinct from what might be obtained, is measured in the first instance by the work done to effect the extraction. Therefore it is appropriate that this appraisal of the ocean's harvest should begin with a survey of the methods of fishing.

Taking a panoramic view of the fishing equipment in use at present, and the way it is used, is a journey through time to observe the evolution of technology. Practically every kind of instrument ever devised for collecting, trapping, or catching fish can be found still in use, somewhere in the world: from primitive baskets for collecting shells, to spears, harpoons, and scoops, to dip nets and cast nets, to beach seines, Danish seines, and otter trawls, to steered midwater trawls. They are used by fishermen walking the shore, or standing on rocks, or diving to collect by hand or to drive fish into a net, or working from canoes, small boats, trawlers, and factory ships. In order to work, fishermen make use, not only of a platform, namely a boat, but of equipment for navigation, for finding fish, for handling and keeping the catch, and for communication.

The equipment (or "gears") used for fishing are divided into three classes. Passive gears are those to which fish are brought by currents, or by their own movements, or by attraction (by light or bait). They include pots and traps, gill nets and drift nets. Active gears are those that move and seek out the fish: among them, seines, scoop nets and cast nets, trawls and dredges. The third class covers equipment directed against individual fish, including spears and harpoons.

To be a fisherman one must be a practical ecologist. One must know the distribution of each target species: where it will be at particular times, and when it will be at particular places. One must know something of its life cycle. And one must know its behavior, in feeding, in moving about its ground and between grounds, in reproducing, and in schooling or remaining solitary. This knowledge relates to what we call "fishable stock", that is fish of a size and kind to be caught; but for some

Opposite. Silhouetted against a Sri Lankan sunset, stilt fishermen perch precariously above a surging tide. From a global perspective, subsistence fishing techniques such as this have a markedly different impact on the total ocean resource than other techniques such as industrial or recreational fishing. The former has more in common with natural predation than large-scale harvesting.

Australian Picture Library

species it is necessary also to know something about the juvenile stock so as to avoid fishing on nursery grounds.

In primitive fisheries this knowledge was obtained by direct observation and applied by reading natural signs, such as weather, sea conditions, and bird behavior. In modern fisheries this knowledge is made more particular and precise by science and technology, and its application is simplified by the use of instruments. But the strategy is unchanged: know the grounds; be on the grounds when the target species will be there at a fishable stage; locate the concentrations of fishable stock; operate the equipment that will take the fish.

KINDS OF FISHERIES

Individual fisheries are named for the species taken (such as herring or salmon); for the location of the grounds (such as Newfoundland Banks); for the kind of equipment used (such as drift net or longline); or for a combination of two or all three of these descriptors. For an appreciation of world fisheries, however, it is better to locate these individual names within a generalized classification based on dominant ecological characteristics of individual species which largely dictate the methods that can be used. Thus we distinguish surface (or pelagic), midwater, and bottom

Fresh eels for sale at the Tsukiji Fish Market, Tokyo. Salted, smoked or pickled, eels have been considered a delicacy since ancient times in both Western and Oriental cultures.

A community effort—fishermen hauling in a beach seine in Kerala, India. Typically a beach seine is towed out in a loop beyond the waves and back to the beach using a canoe or small boat. The net is then brought ashore by pulling evenly from both ends.

(or demersal) fisheries. Each of these classes is divided into those of the shoreline, those of the continental shelf, and those of the deep oceanic waters, except that there cannot be a shoreline midwater fishery.

There is one further complication in this picture of fisheries, namely the distinctions between industrial, artisanal, subsistence, and recreational fisheries. Sport fishing is engaged in chiefly for recreation, even if its catch is taken home to eat, but this is not generally a sustained activity and the catch is not something on which any family depends. Subsistence fishing, in contrast, is a sustained activity carried out in order to obtain food for the fisherman and his family, and perhaps for other members of the community. All these people depend upon this activity, and any sale of a subsistence catch is fortuitous. Fisheries like this still operate in parts of the world, especially from island communities.

A fishing unit is a combination of boat, equipment, and manpower which can operate independently. Artisanal and industrial fisheries differ in the size and ownership of their fishing units. They both deliver their catch to markets, for the fishermen of these fisheries "don't fish for food, they fish for money"; but artisanal fishing is labor intensive, contrasting with the capital-intensive character of industrial fishing. The boat of an artisanal fishing unit is small, generally the equipment is simple and mostly it is hand operated. As

the unit is owned by the person who works it, its operations are carried out pretty much according to the disposition of the owner–operator.

Although the catch is obtained with a view to cash sale, its disposal is arranged through loose arrangements with middlemen and local buyers. In the past most artisanal fishermen were in some sort of bondage to middlemen, who provided credit, sold to fishermen the nets and other materials they required, and fixed the prices at which they took the catch. This situation persists in some places, but with the spread of literacy, and market information available through radio and even television, the hold of these middlemen over artisanal fishermen is steadily being eroded.

Industrial fishing is generally highly organized in terms of the fishing operations, disposal of the catch, and relations with processing plants and markets. The conduct of the fishing operations is dictated largely by the timetable of markets or the production schedules of processing establishments. The fishing units of industrial fisheries are highly modernized in motive power, navigational instruments, fish-finding equipment, winches, communications equipment, and the machinery needed to handle, store, and process the catch. The boats vary greatly in size, from 15-meter (50-foot) shrimp trawlers to great factory ships more than 100 meters (330 feet) long.

THE GLOBAL CATCH

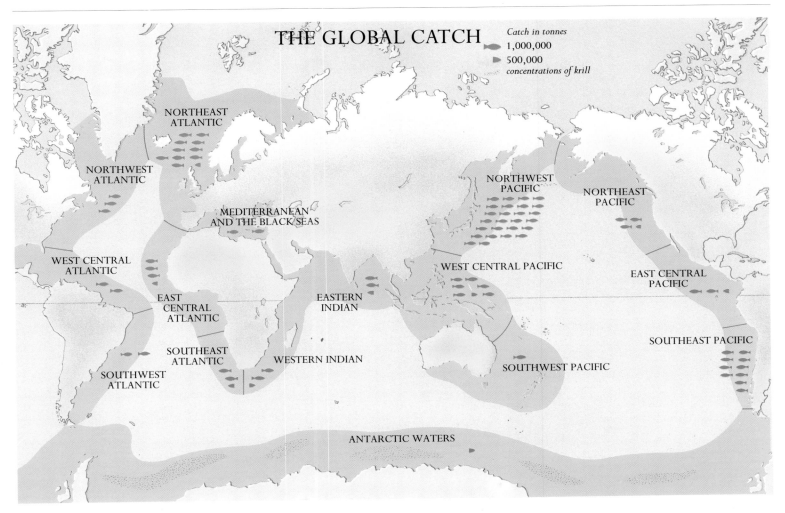

Catch in tonnes
1,000,000
500,000
concentrations of krill

The total global catch of fish, crustaceans, and mollusks (based on Food and Agriculture Organization data), here divided according to the established marine fishing areas, clearly indicates the importance of the Pacific Ocean in world fishing.

THE WORLD CATCH

The species that constitute the living resources of the sea are numerous and varied. The Food and Agriculture Organization's Yearbook of Fisheries Statistics has a list of more than a thousand species of plants and animals "taken for all purposes . . . except recreational" from "inland fresh and brackish water areas and [from] inshore, offshore and highsea fishing areas".

Almost 90 percent of the total reported catch is taken from marine areas. Of the 1987 reported catch of 80,501,200 tonnes, more than 61 percent came from the Pacific Ocean. While some part of the catch is accurately weighed and reported (this is true more especially of the industrial catch), the reports of artisanal and subsistence catch can only be estimates. It may be that exaggerations are cancelled out by omissions, leaving the total more or less correct, but all the figures should be treated as approximations rather than absolutes.

The variation in the catch from area to area within each ocean is a result of important ecological phenomena. The size of the stocks of fish from which the catches are taken depends on the food supply available to each stock, and at base that is the production of phytoplankton. The variation in the richness of phytoplankton

depends upon the supply of nutrients to the places where the phytoplankton can live—notably the surface waters through which sunlight can penetrate.

The reserves of nutrients lie on the sea floor, an accumulation and breakdown of excrement and dead bodies cascaded down from overlying water. When material from these reserves is brought to the surface, as happens where deep-to-surface currents flow, there can be rich phytoplankton growth. It is this upwelling and the associated primary production that accounts for the very great stock of Peruvian anchovetta, from which as much as 10 million tonnes of catch has been taken in a single year.

USING THE CATCH

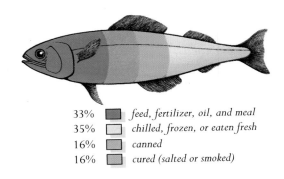

33%	feed, fertilizer, oil, and meal
35%	chilled, frozen, or eaten fresh
16%	canned
16%	cured (salted or smoked)

The marine catch is used in a multiplicity of ways, depending on species, location, and local demand.

Jeff Rotman

Doug Allen/Oxford Scientific Films

Top. *More than 10,000 dollars worth of frozen tuna awaits sale and processing at the Tsukiji Fish Market in Tokyo, Japan.*

Bottom. *Hauling in a cod trawl off Lofoten Island, Norway. Of the many cod species,* Gadus morhua *was the species most commonly fished in the north Atlantic until the fishery collapsed in 1989–90. This collapse was probably partly due to the poor survival of young* Gadus *following the 1983 El Niño year.*

Opposite. *Wrestling a half-tonne monster aboard. The bluefin tuna is one of the biggest as well as one of the fastest of all fishes. Some old fish may reach three meters (10 feet) in length and weigh 1,000 kilograms (2,200 pounds) or more.*

Similar upwellings operate in other parts of the world, but phytoplankton growth is low in great areas of the ocean. In fact, most of the world's catch is taken from only the margins of the seas and continental shelves, many parts of which are narrow and almost desertic in character. The catch from the high seas, chiefly of the highly migratory species—tunas and billfish—is only a very small proportion of the total. The catch of whales and other marine mammals is, of course, now very low.

It will now be clear that the comparison with agricultural production made at the beginning of this discussion is not as clear-cut as it seems. In the first place, comparison should be made between production from truly productive areas on both sides. Secondly, while some part of the total marine catch, probably less than 5 percent, is the product of aquaculture, the bulk is the result of harvesting. Thirdly, the efficiency of harvesting is still constrained by inadequate administration.

THE PROSPECTS

Pessimistic forecasts of the future of the oceans are commonplace. One frequently hears warnings that the resources of the seas are being so heavily overfished that they are likely to be destroyed; that development of coastal areas, with destruction of mangrove stands, reclamation of shallow-water areas, and building of marinas and other structures, is destroying the habitat of valuable species of fish, crustaceans, and mollusks; that pollution is destroying habitats and organisms. And it is often declared that the result of all these activities has already reduced the yield from the seas.

However, this last claim is not supported by sound evidence. Over the past 10 years the total reported catch has increased slightly, and the production achieved by the methods in use at present could be increased by rational management of fishing. Many stocks have been overfished and the catch from them is less than it might be. But if in each case procedures for monitoring the resource could be adopted, and if each year the catch could be limited appropriately to the stock-of-the-year, catch rates could be held at a level which, on average, would be higher than the levels of the past. Thus the total world catch could be increased just from the resources at present exploited.

Then there are the possibilities of extending the range of organisms to be caught. In this category could be included the use of the by-catch of species not sought in shrimp and other fisheries, which for the most part are thrown back dead into the sea. Making use of the by-catch could be part of a set of inter-ventionist techniques in the ecosystems from which catches are taken at present. These might include selective fishing of certain species in order to reduce predation on juvenile stages of sought-after species, and the plan referred to as "ocean ranching" in which broodstock will be held in breeding areas and the offspring fed and cared for until they are of a size to survive upon release into the open ocean where they would find naturally provided food. And, of course, there are the possibilities of aquaculture.

Just how far these possibilities can be realized will be dependent, in part, upon the scope and intensity of the forces working against improved fish harvests. But it is important to take an informed view of those forces. For example, not all the materials deposited in the seas are noxious and damaging. A substantial proportion of the organic materials, particularly domestic sewage, can be rated as a return of nutrients to the sea. Similarly, not all fisheries are conducted in a highly competitive, prone-to-overfish style. Answers to problems in these two areas are not to be found in technology, nor even in natural sciences, but in social sciences—even ethics.

G.L. KESTEVEN

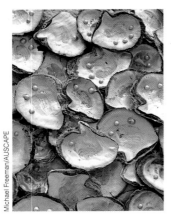

Only a few species of oysters in the genus Pinctada *produce pearls, which are formed when some small object such as a grain of sand strays into the oyster's shell. The oyster envelops the object in a coating of nacre, which consists mainly of calcium carbonate.*

Opposite. *The ocean's bounty on display at the Pike Market, Seattle. Ranging from oysters, prawns, and crabs to tuna and other deep-sea fish, the world's total seafood catch exceeds 80 million tonnes per year.*

Oysters have been farmed for centuries using a variety of methods. These oyster farming boats in western Spain have their valuable crop growing on ropes suspended from the outriggers.

FARMING THE SEA

The farming of fish and shellfish, popularly known as aquaculture, has been undertaken for centuries. As a means of animal production aquaculture has much in common with agriculture, given that the animals have aquatic rather than terrestrial life histories. In both cases the animals are cultured at much higher population densities than those that occur in natural habitats; they require optimal growing conditions including plentiful nutrition; and they are harvested from small enclosed areas or culture containers rather than from open waterways.

Species of fish and shellfish suitable for aquaculture occur in many varieties, sizes, and shapes. The most common group of farmed animals is the molluskan shellfish. These include the bivalve mollusks (which have twin shells) such as edible oysters, mussels, scallops, clams, and pearl oysters; and the monovalves (with single shells), principally the abalones. Prawns and shrimps, and to a lesser extent crabs and lobsters, are typical crustacean aquaculture species. Many types of fish are grown, such as salmon, trout, carp, and tilapia.

Aquaculture species are successfully farmed in diverse tropical, subtropical, and temperate regions around the world. As we approach the end of the twentieth century, aquaculture is a boom industry.

Fish and shellfish are farmed in many ways, and four examples from different parts of the world are outlined here. A traditional example is the Asian village pond or dam stocked with tropical fish. These systems are community based, require water of only moderate quality, and often make efficient use of diverse nutritional sources including waste materials. The aquaculture species may be harvested at intervals from six months to two years, and are likely to be consumed locally at low production cost.

A more technologically based example is the oyster farm in subtropical and temperate estuaries of Australia or New Zealand. These farms are deliberately located in areas of high water quality and high natural abundance of microscopic algae (or microalgae), the oysters' food source. Because the food costs nothing, the farms are often economically attractive. The animals are typically cultured in trays (or on sticks), graded at regular intervals to ensure vigorous growth rates, and harvested after 18 months to three years of farming.

Land-based ponds for prawns in Japan and Taiwan vary from simple low-cost operations to sophisticated systems. The quality of the bottom sediment is particularly important for successful culture. In the intensive feeding style all the nutrients are provided, usually in the form of pellets. The diets are formulated according to the same principles as for fish. In the extensive feeding style a natural bloom of microalgae, zooplankton, and other small members of the food chain is allowed to develop. These small food items or their detritus form the food source for the prawns. In efficiently run tropical farms it is possible to harvest two crops of prawns per annum at moderate production cost.

Marine farms for fish pioneered in Norway and Scotland probably represent the most technologically advanced type of aquaculture grow-out activity. The fish are contained within pens or cages where their stocking density is carefully controlled. If the quality of the water or underlying sediment deteriorates, the pens are moved to a better area. The fish are given pelletized feeds with a defined content of protein, carbohydrate, lipid (including polyunsaturated fatty acids), vitamins, and minerals. The pellets may also have additives such as pigments to color the fish flesh or antibiotics to control possible disease outbreaks. Automated devices can be used to deliver the feeds at specific times in specific quantities. The fish are typically harvested at 18 months to two years of age.

THE FUTURE OF AQUACULTURE

A promising future awaits fish and shellfish farming. It is predicted that in some parts of the world aquaculture production will rival that of fisheries by the turn of the century. Certainly aquaculture is viewed by many fisheries scientists as an essential supplement to the fishing industry. To ensure that the prospects for aquaculture are maximized, substantial biotechnological research is being undertaken on the current range of species. Many new species are being evaluated for their potential. Marketing aspects are also receiving considerable attention from researchers, especially with regard to nutritional value and longer shelf-life.

The major concern for aquaculture is environmental pollution. Our world is increasingly at risk from the exotic materials accumulating in our waterways. Already there have been instances when pesticides and heavy metals (such as nickel, copper, and tributyl tin) have entered hatcheries via the intake sea water and profoundly affected larval survival. Growing animals have been weakened or disfigured, and occasionally killed, by toxic materials. Often their sources have been accidental spills or sediments which have become highly concentrated over the decades.

More sinister is the accumulation of pollutants by fish and shellfish which otherwise appear normal to the human eye and nose. The range of possible pollutants includes heavy metals, pesticides, organochlorines, and disease-causing bacteria and viruses from sewage. The consumer may be the unwitting recipient of these compounds, and suffer acute or chronic illness as a result.

Environmental pollution is of concern not just to aquaculture, but to the capture fisheries and to aquatic life in general. In the next decade society must work diligently to devise and implement strategies to control pollution. The methods will need to be simple and inexpensive so that they can also be applied in developing nations.

CHRISTIAN D. GARLAND

Brian Parker/Tom Stack & Associates

A BOOMING INDUSTRY

CHRISTIAN D. GARLAND

There are many advantages in farming fish and shellfish compared to catching the animals from the wild. It is possible to farm some species all year round and thus not be subject to seasonal variation. Individual animals of reasonably uniform size, weight, and color can be produced to suit the requirements of the retailer or consumer. Animals can also be supplied at regular times and in predictable quantities. All these considerations are of great commercial significance, and they contribute to the increasing acceptance and popularity of aquaculture products.

Center. *The ideal fish for aquaculture has not yet been found, but one breed that has attracted wide attention are members of the fresh water genus* Tilapia, *originally mainly African in distribution. Once described as combining an elephant's appetite with a rabbit's fecundity, these fish can tolerate wide fluctuations in temperature, salinity, and water quality, and are mainly vegetarian in diet. Here they school in an aquaculture tank in Hawaii.*

Several criteria must be satisfied for an aquaculture species to be farmed successfully in large numbers. Firstly, the developmental phases of the animal should be understood. Then the capacity to provide optimal growing conditions (such as oxygen level, temperature, salinity, stocking density, nutritional supply, and lack of predators) is important. Knowledge of techniques to maintain basic hygiene and control disease outbreaks is essential at all times. Opportunities for rapid spread of disease are an unavoidable consequence of the crowded growing conditions associated with aquaculture systems. This is similar to the spread of infectious disease among intensively reared pigs, sheep, and chickens.

Over many years, and particularly in the twentieth century, hundreds of aquatic animals have been

Greg Vaughn/Tom Stack & Associates

PRODUCTION STAGES IN TYPICAL AQUACULTURE OPERATIONS

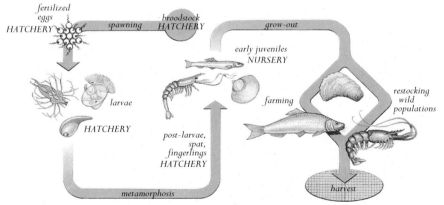

Modern aquaculture farms monitor and control their stock at all stages of production, from spawning to harvest. Good management of the hatchery and nursery stages is crucial to a successful industry.

evaluated for their aquaculture potential. Few have been successful, because of the complex and changeable web of biological, industrial, and marketing factors that impinges on any aquaculture venture. The farming stage of aquaculture is the most time-consuming. It requires a substantial input of technical expertise and capital equipment. A large supply of artificial food may also be needed for some species. But there is more to a successful modern aquaculture venture than the farming stage. After harvest, the processing, post-

harvest handling, and marketing stages must be undertaken. Beforehand, there are the hatchery and nursery phases which provide the juvenile stock.

HATCHERIES AND NURSERIES

Traditionally fish and shellfish farmers obtained their juvenile stock from natural waterways. At the appropriate season bivalve spat, for example, were collected on sticks, or fish fingerlings were trapped in nets. However, these traditional methods have become increasingly unpredictable, often failing to yield the full quantity of young animals that is required by farmers.

Hatcheries, now the cornerstone of many communal and commercial aquaculture industries, were developed in the late nineteenth and twentieth centuries, often in response to the dwindling natural supplies of juveniles. Trout hatcheries were among the first. A modern aquaculture hatchery is a technologically advanced and compact operation. It is usually located on a few hectares of coastal land with access to high-quality sea water. Many thousand liters of water are likely to be needed every day.

Like terrestrial farming, the successful rearing of animals in hatcheries is no accident of fate. Many complex, skilled activities are undertaken there. Firstly,

the breeding animals must be conditioned so that their gametes are abundant and healthy, and bear the appropriate genetic characteristics. The animals are then induced to spawn their gametes, which are collected separately (when possible) by technical staff. The gametes are mixed to achieve fertilization. The young larvae hatch, usually within one to two days.

Larvae are typically reared in tanks containing still or slowly replaced sea water. The culture conditions are carefully controlled, especially with respect to temperature, dissolved gases, pH levels, and stocking density. General hygiene is critical: infectious disease can spread rapidly and tens of millions of young animals may be put at risk.

Since natural sea water contains insufficient items for mass-cultured larvae, the food supply must also be carefully managed. Microalgae are the preferred food of bivalve larvae and spat, and of the early larvae of many fish and crustacean species. Consequently the mass culture of microalgae is often required. As the fish and crustacean larvae develop, they prefer larger live animal food items such as rotifers, copepods, and brine shrimps (which feed on microalgae) so these food sources must also be mass-cultured in the hatchery. Alternatively, the older larvae may take artificial feed such as pellets, but these have been developed for relatively few species.

At regular intervals during the hatchery phase, technical staff catch larvae on sieves and examine them microscopically to check their size, shape, mobility, feeding habits, and organ development. Between one week and several months, and after one or more larval cycles, the animals are ready to metamorphose. Special habitat conditions are then provided so that, over several days, they can transform into juveniles.

The juvenile animals are usually produced in batches, and annual production by some hatcheries can amount to tens of millions of fish and crustaceans or thousands of millions of bivalve mollusk seed. These early juveniles may be despatched immediately to farmers or they may be acclimatized for several weeks in a nursery before being despatched. In either case, an efficiently operated modern hatchery or nursery can supply all the farmers located along several hundred kilometers of coastline—no small achievement! ●

These rocky shore mussels Mytilus *are commonly farmed. Other molluskan shellfish, such as edible oysters, scallops, clams, pearl oysters, and abalones, are also suitable for aquaculture.*

Salmon farming in the Outer Hebrides. Some farms collect their own eggs and raise the fish right through to maturity; others buy juveniles from a hatchery.

MINERALS AND ENERGY

For centuries, people have extracted useful minerals and energy from the sea. The most abundant of the ocean's mineral resources are its salts. On average, the salt content of the oceans is about 3.5 percent by weight, although it varies with the rates of evaporation and precipitation at the surface, with the freezing and thawing of sea ice at high latitudes, and nearshore with local runoff of fresh water from the land. Some 86 percent of the dissolved salt is sodium chloride, but the salts of magnesium, calcium, and potassium, including sulfates, carbonates, and bromides, are economically and nutritionally important minor constituents of sea salt. Salt ponds, which sea water enters at high tide and which are then closed so that the water will evaporate and the salt precipitate, are common where settlement has occurred in coastal regions. International trade in salt has affected history the world over.

The oceans and their beaches have long been exploited as a source of building materials. The rocks, sand, and gravel that have been winnowed, sorted, and concentrated by waves and currents are used worldwide for the construction of roads and buildings. Offshore from today's beaches are fossil beaches, the remnants of earlier geological times when sea level stood lower. These too preserve valuable resources, including submarine placer deposits of gold and other heavy metals, and along the Namibian coast of southwest Africa, diamonds, which are extracted by dredging at the mouth of the Orange River.

METALS AND MINERALS

In recent times, attention has been paid to the minor dissolved constituents of sea water. Periodically, we read of the large amounts of precious metals—gold, silver, copper—that are dissolved in, say, a cubic kilometer of sea water. But a cubic kilometer is a great deal of water, and extracting trace metals that occur in nanograms per kilogram of water (a nanogram is one-billionth of a gram) is a costly and uneconomic process.

Instead of trying to extract dissolved metals directly from sea water, it is possible to recover them after time and the ocean have concentrated them in bottom deposits. Perhaps the best known of these are the manganese nodules found in the Pacific, south of Hawaii and northeast of Tahiti. These nodules contain significant amounts of manganese, iron, nickel, and copper, and minor amounts of cobalt, chromium, tin, and other metals. The world mining industry has expressed interest in extracting the nodules, but refining them raises a number of

OCEAN TECHNOLOGY

Conceptually, one of the simplest of techniques to generate power from tides involves damming a suitable pool: adjustable gates allow the incoming tide to enter freely and flood the pool, but are then closed to force outgoing water to channel through a turbine as the tide ebbs.

The wave-contouring raft exploits the energy of surface waves by linking hinged floating plates with hydraulic rams. As wave action flexes the plates up and down, fluid pressure inside the cylinders increases, forcing the fluid through small turbines to generate electricity.

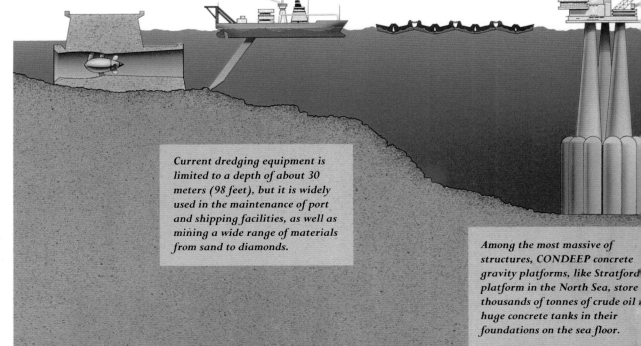

Pressure, darkness, stability, and corrosion define the constraints within which ocean technology and engineering operate, whether the task is mining the seabed, drilling for oil, or general maintenance, exploration, and resource development. A variety of techniques is used to render ocean installations operationally secure. Also, the eternally restless sea could supply vast quantities of electric power if some practical means of harnessing it could be found. Many conceptual methods have been proposed, at least three of which show special promise if the formidable problems of funding them could be overcome.

Current dredging equipment is limited to a depth of about 30 meters (98 feet), but it is widely used in the maintenance of port and shipping facilities, as well as mining a wide range of materials from sand to diamonds.

Among the most massive of structures, CONDEEP concrete gravity platforms, like Stratford platform in the North Sea, store thousands of tonnes of crude oil in huge concrete tanks in their foundations on the sea floor.

environmental, economic, and technical questions that have, thus far, kept them on the bottom of the ocean.

Oceanic mineral deposits result from the slow precipitation of dissolved metals from sea water or by the accumulation of tiny metallic particles carried in bottom currents. As a result, minerals form crusts that cover the sea floor in many regions. Of special interest are the phosphorite deposits that precipitate in areas of high biological productivity, such as areas of coastal upwelling. These deposits represent an economically feasible source of phosphorus, whose main use is as a fertilizer. To date, they have not been exploited as there are many land deposits that are easily mined.

In the active centers of sea-floor spreading, ore deposits are precipitating at much faster rates. Where hot vents bring mineral-rich water from the interior of the Earth and eject it into the cold deep water of the ocean floor, the heated water cools rapidly and the minerals, including many metals and rare elements, are deposited as polymetallic sulfides near the vents. The mining industry is actively investigating the possibility of recovering these deposits. A major difficulty lies in developing techniques to extract the metals that are bound in very complex chemical compounds.

OIL AND GAS

While the salt trade drove the politics of global wars and alliances in earlier times, the most influential economic resources of the ocean today are oil and natural gas. From Indonesia to the North Sea, from the Gulf of Mexico to the Sea of Okhotsk, and from southern California to the Persian Gulf, the recovery of oil and gas from offshore deposits is perhaps the single most important driving force in the global political arena. The value of offshore oil and gas deposits, worldwide, is still controversial, but new fields are being found, and as we learn more about the ocean floor, the discoveries are bound to continue. The production and transportation of offshore oil and gas are major features of the international economy.

Recovery of these resources is a subject of international debate, as offshore drilling accidents have damaged the ocean, at least temporarily. While one of the great strengths of the global ocean is its ability to absorb material, whether it comes from natural or human sources, the better alternative is to extract resources without affecting either the ocean or its inhabitants. The technology used in the offshore oil and gas industry is some of the most advanced in the world.

In parts of the Pacific Ocean, manganese nodules litter the deep-ocean floor. Formed over millions of years by processes still not entirely understood, these lumps of raw metal contain iron, nickel, and copper as well as manganese. Engineering difficulties as well as legal complexities (who exactly owns them?) impede exploitation of this potentially very valuable resource.

Some offshore oil-drilling rigs are guyed, like this platform, called Block 480, in the Gulf of Mexico. It is pinned to the sea floor with a spoke-like array of 900-meter (2,950-feet) cables.

In the tropics a temperature differential of around 20°C (68°F) exists between the ocean surface and depths greater than 1,000 meters (3,280 feet), a difference that can be exploited by OTEC (Ocean Thermal Energy Conversion) devices that function very like a refrigerator in reverse. The warm surface water vaporizes ammonia, the resulting expansion driving a turbine. Then the ammonia is recondensed by cold water drawn up from the depths.

Pilot projects have demonstrated the feasibility of using mechanical "harvesters", remote-controlled from a surface vessel, to gather manganese nodules from the deep seabed, but costs are high and commercial applications only a distant prospect.

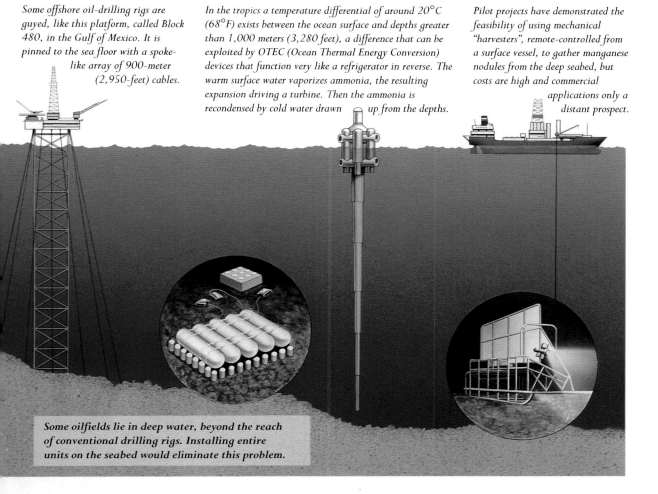

Some oilfields lie in deep water, beyond the reach of conventional drilling rigs. Installing entire units on the seabed would eliminate this problem.

Salt-producing industries such as this one at Lake Magadi, Kenya, most commonly extract their product by solar evaporation of sea water or inland brines. Once the salt has crystallized in shallow pans, it is washed with a saturated brine solution, rewashed with fresh water, dried, and sold.

Opposite. Vast reserves of oil and natural gas lie under the seabed in various parts of the world. Some are under shallow offshore waters, but others lie beneath the deep-sea bed, where extraction requires an enormous investment in technology and capital. Here, a self-contained offshore oil-drilling rig nears completion before being towed to a location in the North Sea. Almost a million tonnes in weight, some of these floating cities are among the largest objects ever moved by humans.

SOLAR ENERGY FROM THE SEA

Although the recovery of the mineral resources of the ocean is an ancient human endeavor, the recovery of usable energy from the ocean has a number of modern and futuristic dimensions. The ocean's energy comes from the sun, which moves the atmosphere and the oceanic currents, and from the gravitational energy of the tides. Studies of the ocean floor have revealed very large geothermal resources as well, but the recovery of geothermal energy is in its infancy even on land, and ocean floor geothermal power is years away.

The solar energy of the ocean is stored in the surface layer—the upper 200 meters (650 feet) or so. The simplest form of energy to extract is that contained in the energy contrast between the surface and the deep layers of the ocean. Just as a refrigerator requires energy to maintain a temperature difference between its interior and exterior, energy is available wherever there is a temperature contrast in the ocean. The greatest contrast between the surface and deep layers occurs in the equatorial ocean, and between latitudes 20°N and 20°S the difference is large enough to recover energy through a process known as Ocean Thermal Energy Conversion (OTEC). The process has proven economically feasible at pilot plants such as the one at Keahole Point on the island of Hawaii. An improvement in the technology has also produced fresh water as a by-product of the energy conversion process—a valuable resource for island nations in many parts of the world.

The second major source of solar energy in the ocean is from currents. These great rivers in the sea flow constantly and along well-known routes. It may one day be possible to produce useful energy by mooring turbines in the major fast-moving currents such as the Kuroshio, the Gulf Stream, and the East Australia Current.

Obviously, transmitting energy produced in mid-ocean, or for that matter on sparsely inhabited oceanic islands, through the OTEC process may introduce significant technological problems. But it is also possible to transmit energy in indirect ways. We now ship bauxite from the Caribbean to Scandinavia where cheap hydroelectric power makes the refining of aluminum economical. If the energy of the ocean is used near the production site to effect an energy-intensive process, such as bauxite refining, and the refined product is shipped, the energy is transmitted at a greatly reduced cost. Still, direct transmission is possible. Iceland is currently studying the possibility of transmitting inexpensive geothermal and hydroelectric power to Ireland and Britain by submarine cable, and similar studies have been carried out in other parts of the world ocean.

WAVE AND TIDAL ENERGY

The atmosphere moves the ocean to create currents, but it also produces surface and internal waves in the water. Although these waves contain a great deal of energy, we have yet to develop a means of extracting it in an economical and environmentally safe manner. At present, wave energy remains a largely destructive rather than constructive resource.

The second major source of oceanic energy is the gravitational energy embodied in the tides, and this energy has been used in many ways. The simplest use is, perhaps, in salt ponds. The tides also lift ships into and out of drydocks around the world every day. Sea water flooding low-lying land in Maine and the Low Countries is drained back to the sea through turbines to produce hydroelectric power. Narrow straits such as the Bering Strait and the entrance to the Bay of Fundy may one day produce energy from tidal currents.

Many sources of oceanic energy are available, but we have learned to use only a few. Unlike the mineral resources of the sea, its energy is renewed continuously by the sun and by the gravitational attraction of the moon, sun, and the other celestial bodies. ■

JAMES C. KELLEY

10 PLANNING FOR THE FUTURE

ALEC C. BROWN, ANGELA MARIA IVANOVICI, AND VICTOR PRESCOTT

T he oceans form a link between all who inhabit our planet, and their well-being is a matter of common concern. The oceans challenge science and technology, national institutions, and international resolve. Who owns the oceans and what rights do individual nations have to the resources within them? How serious is the threat of oceanic pollution and what is being done about it? Has human intervention already destroyed the balance of nature? How is the oceanic environment being protected and is the current level of protection in need of revision? The answers to these questions will help to shape the future of the world's oceans.

Opposite. The dark flicker of an outrigger canoe from below evokes an image used on tourist brochures throughout the Western world as representing the ultimate holiday destination. Matchless beauty and wonder lie within a few meters of the surface along the coral reefs of the world, a priceless heritage for future generations.

THE LAW OF THE SEA

Before the Second World War most maritime claims were to territorial seas three nautical miles wide: a distance about the range of cannon fire in the eighteenth century. After the war, however, the extent of claims varied. Two United Nations conferences, in 1958 and 1960, failed to solve the problems resulting from this ad hoc development.

In December 1982, 119 delegations signed the Law of the Sea Convention which had been hammered out in the previous eight years at the Third United Nations Conference on the Law of the Sea. That Convention will come into force after 60 countries have ratified it and a year has elapsed after the sixtieth ratification. At the beginning of 1992, 51 countries had ratified the Convention. Even though it is not yet in force, many countries appear to be applying the Convention's rules, and so it is appropriate to describe them here.

THE FOUR ZONES

Under the 1982 Convention, coastal states are entitled to claim four main zones. Proceeding seaward from the low-water line, the first zone is called the territorial sea. With one exception the coastal state has complete authority over the territorial sea, and over the air-space above it and the seabed beneath it. The solitary exception is that foreign vessels are entitled to exercise the right of innocent passage through territorial seas. Passage is deemed to be innocent if it is not prejudicial to the peace, good order, and security of the coastal state. The maximum permitted width of the territorial sea is 12 nautical miles from the baseline used by the coastal state. Normally that baseline is a low-water line. One nautical mile is equal to one second of latitude, which is 1,852 meters (6,075 feet).

The contiguous zone lies beyond the territorial sea, and in it the coastal state may exercise controls necessary to prevent infringements of its customs, fiscal, immigration, and sanitary laws. The outer limit of the contiguous zone may not be more than 24 nautical miles from the baseline. The next zone is the exclusive economic zone. This zone may have a maximum width of 200 nautical miles measured from the baseline. In it the coastal state has the sovereign right to manage and exploit the living and non-living resources of the waters and the seabed. This right includes the generation of power from waves or tides. The coastal state also has the sole right to establish artificial islands in the zone, and jurisdiction over research and protection of the marine environment. If artificial islands or other installations are established, the coastal state may create a safety zone of 500 meters (1,650 feet) around them, within which it can make rules for navigation.

If the continental margin, which consists usually of a continental shelf, continental slope, and continental rise, extends more than 200 nautical miles from the baseline, the coastal state may claim it. The claim is only to the seabed and is not to the waters above that seabed. On the continental shelf claim, as in the seabed of the exclusive economic zone, the coastal state has the sole right to manage the resources of the seabed. Countries with shelves wider than 200 nautical miles include Australia and the United States.

This schematic representation indicates the full range of national maritime claims as agreed at the 1982 Law of the Sea Convention.

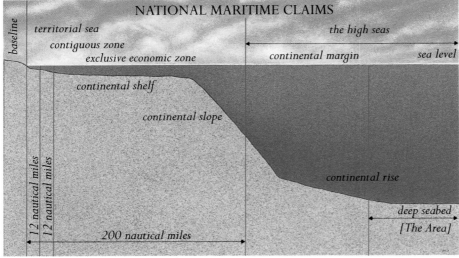

NATIONAL MARITIME CLAIMS

baseline

territorial sea
contiguous zone
exclusive economic zone

the high seas

continental margin   sea level

continental shelf

continental slope

continental rise

12 nautical miles
12 nautical miles

200 nautical miles

deep seabed
[The Area]

There is no country in the world so distant from its neighbors that it can make the full suite of maritime claims without overlapping with similar claims from a neighbor's coast. In such cases the countries involved must agree on how to divide the waters and seabed which they could both claim. At the beginning of 1992 about 125 marine boundary agreements had been concluded. Most of them were in enclosed or semi-enclosed seas such as the Caribbean, the North Sea, the Baltic Sea, the Mediterranean Sea, the Persian Gulf, and the Andaman Sea. In some cases countries have asked the International Court of Justice or an arbitral tribunal to select the appropriate boundary or the means of defining any boundary. Significant cases have included those between France and the United Kingdom over the boundaries in the Channel, and Canada and the United States over the boundaries in the Gulf of Maine.

The pure and bitter grandeur of ice-locked Antarctica conceals vast resources beneath the ice and frigid encircling seas, especially actual and potential food resources in the form of whales, fish, and krill. Yet, because there are no territorial claims in the Antarctic region, wise management of these riches will require international cooperation on a scale now only gradually being realized.

There are two international zones in the oceans. First, the high seas lie beyond national claims to territorial seas and the exclusive economic zone. All states, whether coastal or landlocked, have equal rights to make peaceful use of the high seas. Second, the deep seabed, called The Area in the Convention, is the seabed beyond national claims to the seabed of the exclusive economic zone or the continental margin where it is wider than 200 nautical miles. Where states claim margins beyond 200 nautical miles the waters above that distant margin will be part of the high seas. The Convention lays down detailed rules for mining The Area; an international body provided for in the Convention will oversee any such mining. These rules were developed to control the mining of manganese nodules, which has not, as yet, occurred. Polymetallic deposits within the exclusive economic zones of some states, including the United States, offer more attractive mining ventures beyond the authority of the international body.

David Hiser/The Image Bank

ISLANDS AND ARCHIPELAGOS

Countries can claim maritime zones from islands as well as from the coasts of mainlands. So the United States can claim an exclusive economic zone around the Hawaiian Islands, in the same way that India can around the Nicobars and Britain around the Falkland Islands. There is a rule by which claims to territorial seas only are allowed from rocks. However, it is imprecise because it provides no way of properly distinguishing a small island from a large rock.

Archipelagic states enjoy special provisions under the Convention. If a country is constituted wholly by one or more archipelagos, it can draw straight lines connecting the outermost islands and measure its maritime claims from those straight lines. There are rules which prevent some widely scattered archipelagic states such as Kiribati or some very compact ones such as New Zealand from drawing such lines, but archipelagic baselines have been drawn by countries such as Indonesia, São Tomé and Príncipe, and Fiji.

The only continent around which maritime claims have not been made is Antarctica. The members of the Antarctic Treaty have established conventions to protect marine animals and fish, and to prohibit the mining of resources, such as minerals and fuels on Antarctica and its continental shelf. Provisions have been made for monitoring ecosystems.

VICTOR PRESCOTT

OCEANIC POLLUTION

For most of human history, the sea has been regarded as a vast sink into which anything could be dumped with impunity. Those who thought about it at all, assumed that dissolved substances rapidly became diluted in the sea until their presence could not be detected, while just about everyone believed that sea water killed all "germs".

The single event that changed this attitude and stimulated marine pollution research was the outbreak of Minamata disease on Japan's most southerly island in the 1950s. After many false trails had been followed, the origin of the disease was traced to the release of mercury into Minamata Bay by a local factory—but by then scores of people had died from eating mercury-polluted fish and a still greater number, mostly children, were permanently paralyzed.

Research undertaken since then has shown that virtually all previously held concepts were wrong. Many pollutants do not become progressively diluted but are trapped and even concentrated in marine sediments, from where they may enter the food chain. The pollutant is further concentrated as animal eats animal, the concentration in the final predator—usually a fish, bird, or mammal—being up to a million times what it was in the sea water. Humans, at the top of the food chain, therefore run the greatest risk of all. Once in the sea, pollutants may be converted to even more toxic forms; an example was the conversion of inorganic to organic mercury in Minamata Bay.

Palau from the air. In some parts of the world island archipelagos are shared by two or more nations, raising complex issues of territorial rights over offshore waters. This has consequent implications for conservation and resource management.

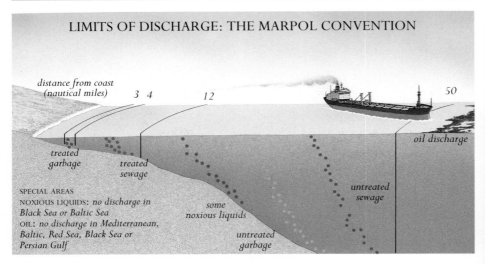

LIMITS OF DISCHARGE: THE MARPOL CONVENTION

distance from coast
(nautical miles)   3 4   12   50

                                                    oil discharge

treated
garbage   treated
               sewage

SPECIAL AREAS
NOXIOUS LIQUIDS: *no discharge in*
Black Sea or Baltic Sea   untreated
OIL: *no discharge in Mediterranean,*   sewage
Baltic, Red Sea, Black Sea or   some
Persian Gulf   noxious liquids

                              untreated
                              garbage

Hans-Jürgen Burkard/Bilderberg

Peter Menzel/Australian Picture Library

Top. *In 1973 the International*
Convention for Prevention of
Pollution from Ships (MARPOL)
was established. It sets minimum
distances from the coastline for the
discharge and dumping of treated
and untreated garbage and sewage,
noxious chemicals, and oil. Its
provisions state that oil cannot be
discharged from tankers in heavily
polluted and vulnerable areas. To
date, the provisions of MARPOL
have not been fully adopted by the
international community.

By 1970, when at last the United Nations began to take an interest in marine pollution, the situation had become very serious indeed. The Mediterranean was dying; the Bosphorus resembled a giant sewer; many European estuaries were completely fouled; cities were having problems disposing of their sewage; and pesticides such as DDT could be found in the tissues of most marine animals, even in the remote Antarctic.

It was also discovered that, far from the sea killing all pathogenic organisms in sewage, the causative agents of such diseases as cholera, typhoid, and hepatitis could survive for long periods in sea water and were accumulated in shellfish such as mussels and oysters. In 1973 the United States Environmental Protection Agency proposed that even the slightest contamination of coastal waters with sewage was unacceptable, although the majority of countries still do not take this advice seriously and there are many areas where sewage contamination makes swimming and shellfish consumption hazardous.

There is no question that we will always have to endure some marine pollution; what we can hope for is to reduce it to an acceptable level, a level which the sea can assimilate without grave ecological effects and without health hazards to humans. Unnecessary pollution cannot be tolerated—and it must be stated that a good deal of marine pollution is quite unnecessary and easily prevented. For example, the mercury spilled by the factory in Minamata Bay was not a waste product but a valuable catalyst: it escaped because of a faulty

countries has been to increase the length of these pipes, this can never be regarded as the final answer; neither can it be considered appropriate to transport sewage sludge offshore and dump it, as is done in New York.

In most countries, a permit is now required for the discharge of an industrial effluent to sea, although in many cases the legislation governing the granting of permits is unnecessarily complex or open to abuse. For many years it was considered that once "safe" discharge levels had been established for one factory, these could be applied to all other factories with similar effluents. This was poor thinking, as clearly a factory discharging into strong currents in the open sea can afford to have higher levels of toxicants than one discharging into an estuary or a sheltered bay, where mixing and turnover of the water body are slow. Also, other discharges into the same area must be taken into account, so that acceptable seawater quality is maintained. The question is, what is acceptable?

For a long time the toxicity of an effluent was tested by discovering the concentration that would kill 50 percent of a given species (usually a fish or a crustacean) in a given period (usually 48 or 96 hours). A "safe" discharge level was then considered to be a tenth, a fiftieth, or a hundredth of that value, depending on the country concerned. Such simplistic principles have now largely been abandoned in favor of much more sensitive sublethal tests, in the knowledge that almost any change in the physiology or behavior of a population will reduce its reproductive potential and is thus likely to eliminate it in the long term. Sublethal testing has reached a high level of sophistication and is constantly being refined. What one would really like to know, of course, is the effect of a pollutant on the whole ecosystem, in order to evaluate the "assimilative capacity" of the system, but this is difficult and remains something for the future.

The concepts of assimilative capacity and seawater quality (rather than effluent quality) have been combined with the notion of "beneficial seawater uses", the idea being that sea water used to cool a power station need not be of very high quality, while at the other end of the scale areas used for aquaculture or recreation must have water of the highest purity. The concept is a useful one, although it has obvious dangers for the marine environment.

An enormous amount of research has been undertaken on marine pollution, and it continues with ever-increasing momentum. The results have already been applied in a practical way, and this, combined with improved river-water quality and reduced pollution from runoff from the land, might be expected to alleviate the situation. So it would were it not for the continuing increase in the world's population. More effort is therefore needed, coupled with greater understanding of marine ecosystems, if the oceans are to survive.

ALEC C. BROWN

recycling plant. When the fault was rectified, not only did the pollution cease but profits increased as the demand for additional mercury decreased.

Pollution by plastic materials, a comparatively recent but now very widespread and severe form of pollution, is largely preventable by public education coupled with appropriate legislation. Most pollution by crude oil is also preventable and in fact has decreased quite dramatically since courts began to impose heavy fines for avoidable spills. Sewage disposal continues to present a problem, although many countries have resolved that no new permits will be granted for the discharge of raw or partially treated sewage to sea. However, a great many pipelines discharging raw sewage remain in operation and while the tendency in developed

Max Tilley/Australian Picture Library

Because birds rely on their feathers for thermal control nearly as much as for flight, even a small amount of drifting oil caught up in the plumage is often fatal, breaking the plumage seal like a pin-prick in a spacesuit. Pneumonia, starvation, or poisoning from oil ingestion soon follows.

Opposite, bottom. *A team works to clean up a polluted beach in Prince William Sound, Alaska, after a major oil spill. Oil spilled on the high seas produces severe pollution damage, but if the slick should be cast up along a coast before it disperses, the results are usually catastrophic, destroying local marine plant and animal communities, sport and commercial fisheries, and aesthetic and recreational amenities.*

Center. *Hundreds of oilwells were systematically fired when the Iraqi army withdrew from Kuwait during the Gulf War in 1991. It took nearly a year merely to extinguish all the fires, much longer to clean up the mess and restore normal production.*

Dead coral. Sometimes thousands of kilometers long, coral reefs are the largest things ever constructed by living organisms. Moreover, they represent sites of animal diversity and complexity rivaling anything else on the planet. Yet coral reefs are also extraordinarily vulnerable to pollution damage from diverse causes ranging from oil spills to chemical runoff from nearby rivers and streams.

PROTECTING THE MARINE ENVIRONMENT

More than 300 marine and estuarine sites around the world have been reserved as protected areas. They are known variously as marine parks, sanctuaries, or reserves; aquatic, fisheries, habitat, wetland, or nature reserves; historic or shipwreck sites. Marine protected areas have been reserved to protect endangered species, marine and estuarine environments, aesthetic and cultural values; to manage and improve fisheries; and to provide opportunities for conservation, education, and scientific or historical research. They vary in size from less than 3 hectares (7½ acres), as for example at the Shiprock Aquatic Reserve in Port Hacking near Sydney, Australia, to the approximately 35 million hectares (87 million acres) of the Great Barrier Reef Marine Park off the Queensland Coast, Australia.

Initially, areas were set aside to protect the beauty of their marine features for tourists. The first marine protected areas, established in the 1930s, were located at the Dry Tortugas off southern Florida in the United States, and Green Island in the Great Barrier Reef. Marine areas have since been reserved in many countries specifically to protect endangered marine species and their habitats. For example, over 19,800 square kilometers (7,640 square miles) in the southwest Pacific Ocean (Lihou Reef and the Coringa–Herald national nature reserves in the western Coral Sea) and the northwest shelf of Australia (Ashmore Reef National Nature Reserve) have been set aside by the Australian government primarily to protect the feeding and breeding habitats of migratory seabirds and turtles.

By the mid-1980s coastal populations of the black cod *Epinephelus daemelli*, a large, sedentary reef fish once common on the southeast Australian coast, had declined to the extent that it was declared a protected species in Australian waters. Its protection has been helped by the establishment in 1987 of the Elizabeth and Middleton Reefs Marine National Nature Reserve (east of the New South Wales coast). Each reef supports a large population of the black cod, and commercial and recreational fishing are restricted within the reserve.

Whales are a significant resource in many protected areas. The St Lawrence River in Canada has been home to the beluga whale *Delphinapterus leucas* since the last ice age. Today only several hundred belugas remain of the 5,000 strong population reported at the beginning of the century. Despite total bans on hunting in recent years, a continuing decline appears to be related to high levels of toxic industrial compounds in the environment. Concern in the community has led to a multimillion dollar Canadian government project to establish a marine park to protect the whale's prime habitat in a 96-kilometer (60-mile) section of the St Lawrence River. Simultaneously, the government will be acting to reduce levels of toxic materials entering the habitat from land-based industries.

Opposite. *Placed on the World Heritage List and gazetted in 1981, Australia's Great Barrier Reef Marine Park sprawls across 35 million hectares (about 87 million acres) of shallow tropical seas. Such activities as littering and spear-fishing with scuba gear are prohibited throughout the park. Though access to some 85 percent of the region is otherwise unrestricted, access to some especially sensitive areas is more rigorously regulated. The park's varied attractions support an enormous tourist industry.*

The Channel Islands National Marine Sanctuary off southern California, established in 1980, is the largest marine protected area in the United States. Many of the world's whale species, such as California gray, humpback, blue, fin, and sei whales, regularly migrate through it. Southern humpback whales swim along the east coast of Australia from Antarctica to breeding grounds in the Coral Sea, and many stop over in Hervey Bay (south of the Great Barrier Reef Marine Park). The Hervey Bay Marine Park was established in 1989, primarily to control a rapidly growing whale-watching industry in the bay and to minimize possible adverse effects of this industry on the whales.

HOW MUCH PROTECTION?

Because many communities have feared the possible or perceived effects of restrictions, most marine protected areas have had to allow continued access to the natural marine resources for recreational and commercial purposes. However, many features for which the areas were protected are beginning to degrade in quality. Lush coral growths in parts of the Florida Reef Tract off Key Largo which were protected in 1960 (John Pennekamp Coral Reef State Park) and 1975 (Key Largo National Marine Sanctuary) have been damaged or have died. Fish and lobsters are becoming harder to catch as numbers decrease. These

declines are attributed to too many visitors and too much pollution resulting from increases in nearby coastal residential developments lacking appropriate controls.

Kelp forests and populations of the black abalone in the Channel Islands National Marine Sanctuary have declined sharply in the past five years. Pressures imposed by continuing human activities are intensifying the effects of natural stresses.

Some marine protected areas do exclude exploitative activities. Studies in the Philippines and New Zealand are finding important differences between populations of commercial and recreational species inside and outside totally protected areas, which raises the possibility of replenishing stocks in exploited waters from those in the protected areas.

The Cape Rodney-to-Okakari Point Marine Reserve (north of Auckland, New Zealand), established in 1975, excludes all forms of fishing and collecting, and access to visitors is limited. Research conducted there by the University of Auckland Marine Biological Station indicates that local snapper and lobster are larger and more numerous inside the reserve than in exploited areas beside it. The reserve was originally considered a threat by the local commercial fishing community, but its beneficial effects are now recognized and the industry contains some of its most ardent defenders against illegal fishing activity.

Many marine mammals and reptiles are at serious risk of extinction, and a few are already extinct, a global problem as shown in this map of the world's oceans and their most endangered inhabitants. Apart from the animals shown, many fish, seabird, and invertebrate species are also threatened.

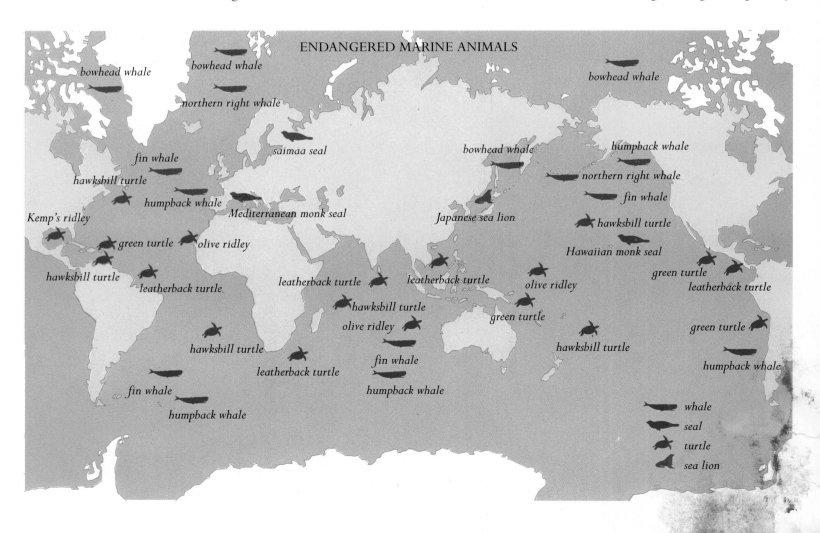

ENDANGERED MARINE ANIMALS

THE GREAT BARRIER REEF MARINE PARK

The Great Barrier Reef Marine Park, which extends for some 2,000 kilometers (1,250 miles) along the northeast coast of Australia, is the world's largest and best-known marine protected area. Inscribed on the World Heritage List in 1981, it includes a vast array of tropical habitats including fringing coastal reefs, coral cays, barrier reefs, and deeper areas between the reefs and offshore. Its four regional management areas are zoned through extensive community consultation and regular review to accommodate commercial and recreational activities.

General use zones (up to 85 percent of the total area) provide for most activities; more protected zones allow only non-exploitative activities such as underwater photography and observation; and reference and research zones (1 percent or less of the total area) totally exclude the public. Significant bird and turtle breeding areas are closed seasonally, as are "reef appreciation areas" and "replenishment areas". Oil drilling and mining, spearfishing with scuba-diving equipment, capture of the potato cod, giant groper, and other nominated species of fish, and littering are totally prohibited throughout the park.

THE FUTURE

Major challenges to the management of this marine park, and many other marine protected areas, include an exploding tourism industry; damage to reefs by crown-of-thorns starfish and other species, and excess nitrogen, phosphorus, and sediments from land-based activities; and the decline of commercial and recreational fisheries.

The survival of healthy oceans to provide a wealth of natural resources now and in the future will depend on extensive protection and integrated management of the marine environment through the reservation of significant or representative coastal and offshore areas. All lovers and users of the sea have a responsibility to promote the protection of our marine heritage by encouraging cooperation within our communities. ▪

ANGELA MARIA IVANOVICI

The beluga whale occurs in the Arctic Ocean and in certain other northern waters. A severe decline in the population inhabiting Canada's Gulf of Saint Lawrence, probably caused by pollution in the Great Lakes, has been countered with concerted local community support, the establishment of a marine park in the beluga's optimum habitat, and vigorous government curbs on pollution along the seaway.

Notes on Contributors

MARGARET ATKINSON
Margaret Atkinson is a research assistant at the University of Sydney, and is currently coordinating a biological survey of the epifauna and algae of rocky reefs in Jervis Bay, New South Wales. She completed a BSc and MSc (Hons) at the University of Auckland before moving to Sydney, Australia, in 1987. Her fascination with the sea and most things marine has resulted in work on various projects at the University of Auckland's marine laboratory at Leigh, the Australian Museum, and the University of Sydney. Much of this field-based work has taken her on many memorable trips to some of the most interesting coastal regions of New Zealand and the east coast of Australia.

MICHEL A. BOUDRIAS
Michel A. Boudrias teaches invertebrate zoology at the University of San Diego. He became interested in marine biology in high school and pursued his interests by working on barnacle functional morphology as an undergraduate at McGill University in Montreal, Canada. His marine biology degree led him to Oregon State University where he studied the feeding and life history of Arctic amphipods. He received his PhD from Scripps Institution of Oceanography, University of California, San Diego. At both Oregon State and Scripps, he became involved in deep-sea photographic analysis and participated in the generation of a biological scale map of the Rose Garden hydrothermal vent. At Scripps, he studied crustacean locomotion, particularly the fluid dynamics and functional morphology of swimming deep-sea amphipods.

ALEC C. BROWN
Alec Brown is Professor of Marine Biology and member of the Marine Biology Research Institute at the University of Cape Town, South Africa. Born in Cape Town, he attended Rhodes University and then worked on water treatment for the CSIR before joining the staff of the University of Cape Town. He was appointed Professor and Head of the Department of Zoology in 1975. He has written five books and over 150 research articles, has worked in Chile and Antarctica, has been visiting Professor to the University of Manchester, and has worked at the University of Cambridge and University College, London, as well as at several marine institutes in the United States. He is a Life Fellow of the University of Cape Town, past president of the Royal Society of South Africa, and holder of the gold medal of the Zoological Society of Southern Africa.

MICHAEL BRYDEN
Michael Bryden has spent the past 26 years dedicated to the study of marine mammals. In 1963–64 he worked as a veterinarian in Tasmania, then spent three years as a biologist with the Antarctic Division of the Australian government, which included 16 months at Macquarie Island studying growth and development of the southern elephant seal. Professor Bryden has held academic posts at the universities of Queensland and Sydney, Australia, and Cornell University in the United States. He has studied the growth and adaptation of seals on four summer research trips to Antarctica and one to the Norwegian Sea, and carried out research on the reproductive biology of cetaceans at the University of Cambridge in 1978 and 1981. In 1988, he took up the Chair of Veterinary Anatomy at the University of Sydney, and began a project to study humpback whales in Hervey Bay, Queensland. He is joint editor of three books, and author or joint author of more than 80 research papers and eight government reports.

M.G. CHAPMAN
M.G. Chapman has a BSc (Hons) degree from the University of Natal and a PhD in marine ecology from the University of Sydney. She is a professional officer at the Institute of Marine Ecology at the University of Sydney, where her work involves experimental ecology and environmental assessment. She is also a writer and scientific advisor for publishing companies.

SYLVIA A. EARLE
Sylvia A. Earle is President and Chief Executive Officer of Deep Ocean Engineering Inc. A marine scientist with a PhD from Duke University, she has held research positions in several universities. In 1982 she co-founded Deep Ocean Engineering Inc. to design, develop, manufacture, and operate equipment in the ocean. She has extensive field experience as a deep-diving aquanaut, has led more than 50 expeditions using a variety of submersibles and equipment, and holds the depth record for solo deep diving. She has written and lectured widely on marine science and technology and is the recipient of numerous prestigious awards to honor her pioneering work. Dr. Earle has recently held the position of chief scientist at the National Oceanic and Atmospheric Administration.

SCOTT C. FRANCE
Scott C. France is a Postdoctoral Fellow at the Woods Hole Oceanographic Institution. His interest in biological oceanography developed during studies at Concordia University in Montreal, Canada. As an undergraduate, he researched the ecology of freshwater copepod populations, and became fascinated with the problem of how deep-sea hydrothermal vent fauna propagate from one location to another. To prepare for graduate studies, he spent a summer in the laboratory of Dr. J. Frederick Grassle at Woods Hole Oceanographic Institution, where he analyzed thousands of photographs of deep-sea hard-bottom communities. He received his PhD from the Scripps Institution of Oceanography, University of California, San Diego. His research at Scripps and Woods Hole concerns the use of genetic population structure analysis for inferring levels of dispersal in deep-sea invertebrates.

CHRISTIAN D. GARLAND
Christian D. Garland is a senior lecturer in the Departments of Geography and Environmental Studies, and of Agricultural Science, University of Tasmania, Australia. He received his doctorate from the University of New South Wales (Sydney) and has worked as a marine scientist in Tasmania for the past 13 years. His initial contact with aquaculture was as a microbiologist investigating disease outbreaks in a commercial oyster hatchery. His interests then broadened to the general importance of bacteria and microalgae to mass-cultured marine animals, fish, and shellfish. Dr. Garland has visited numerous aquaculture ventures in Australia and the Indo-Pacific area, and has seen the industry increase dramatically in scope and financial value since the mid-1980s. He leads a research program on bivalve aquaculture, and is also involved in assessment of environmental pollution in aquatic habitats, especially urbanized estuaries.

STEPHEN GARNETT
Stephen Garnett has always been drawn to seabirds. As a child he worked with the legendary Dom Serventy as he unraveled the story of the short-tailed shearwater in Tasmania, and he has since led or participated in expeditions to study albatrosses, penguins, frigatebirds, and boobies. His research on birds, crocodiles, turtles, and goats has taken him to islands all around the Pacific. More recently he has been involved in the production of a comprehensive handbook that includes all the seabirds of Australia and Antarctica. He is a consultant to the Food and Agriculture Organization, and the Queensland and Northern Territory governments, and is currently working on the conservation biology of threatened parrots in northern Australia.

RICHARD HARBISON
Born in Miami, Florida, Richard Harbison received his AB from Columbia University in 1966 and his PhD from Florida State University in 1971. In 1972 he joined the scientific staff of the Woods Hole Oceanographic Institution, where he is presently a senior scientist. Between 1980 and 1982, he was a principal research scientist at the Australian Institute of Marine Science, and also worked as Director of the Division of Marine Sciences at the Harbor Branch Oceanographic Institution from 1987 to 1989. His primary research involves studying the systematics, physiology, behavior, and distribution of large planktonic animals, such as medusae, siphonophores, ctenophores, salps, and pteropods. Dr. Harbison is mainly interested in animals that inhabit the open sea, but has also studied the gelatinous macroplankton of the Arctic and Antarctic.

ROBERT R. HESSLER
Robert R. Hessler is Professor of Biological Oceanography, Scripps Institution of Oceanography, University of California, San Diego. A childhood fascination with arthropods and fossils led him to study trilobites at the University of Chicago, then to a PhD in paleobiology. He went to Woods Hole Oceanographic Institution in 1960, to collaborate with Howard Sanders in describing the anatomy of the Cephalocarida, the most primitive living crustacean, which he had discovered in 1955. Sanders, a benthic ecologist, began work on the ecology of deep-sea bottoms in 1960, and invited Hessler to join him. Since then he has pursued interests in arthropods and the deep sea concurrently. In 1969, he was invited to join the faculty at Scripps Institution. When hydrothermal vents were discovered, he was part of the first biological expedition in 1979 and has been studying them ever since.

ANGELA MARIA IVANOVICI
Angela Maria Ivanovici is Senior Project Officer with the Australian National Parks and Wildlife Service, with responsibilities for the selection and proclamation of marine and estuarine protected areas. Born in Italy and educated in Australia, she pursued postdoctoral studies in the United States (on a Harkness Fellowship) and the United Kingdom. She is vice-president of the Coast and Wetland Society and secretary to the Australian committee of the International Union for the Conservation of Nature's subcommittee on marine reserves.

JAMES C. KELLEY
James C. Kelley was educated in geology and joined the faculty in oceanography at the University of Washington in Seattle in 1966. His research interests have been in biomathematics and applied statistics in the analysis of marine productivity data measured in coastal upwelling systems. He has led and participated in many research cruises. In 1975 he became Dean of Science and Engineering at San Francisco State University. Since 1986 he has been president of the California Academy of Sciences.

KNOWLES KERRY
Knowles Kerry is a senior research scientist with the Australian Antarctic Division. Since joining the Australian National Antarctic Research Expeditions in 1966 as a biologist, he has wintered on Macquarie Island and spent many summers there and in Antarctica conducting biological research. He has led numerous expeditions to Antarctica. As a permanent member of the scientific staff of the Australian Antarctic Division he established in 1972 the Australian biological research program in Antarctica and in 1979 the marine biology program. He served on the Australian delegation which negotiated the Convention for the Conservation of Antarctic Marine Living Resources and represented Australia at meetings of the scientific committee of the commission. He has recently edited *Antarctic Ecosystems: Ecological Change and Conservation*.

G.L. KESTEVEN

G.L. Kesteven graduated from the University of Sydney, and has worked as a fisheries investigation officer with the New South Wales Department of Fisheries; a fisheries biologist with the CSIR; Deputy Controller of Fisheries in the Department of War Organization of Industry; and an advisor on fisheries to the South Pacific office of UNRRA. In 1947 he joined the Food and Agriculture Organization and was stationed in Singapore, Bangkok, and Rome until 1960 when he returned to Australia to become assistant chief of the CSIRO Division of Fisheries and Oceanography. In 1967 he rejoined FAO, undertaking fieldwork in Mexico, Peru, and other Latin American countries. He has since retired, but continues to act as a consultant to fishery companies.

M.J. KINGSFORD

M.J. Kingsford was born in Hastings, New Zealand. After a first degree at Canterbury University, he undertook research for MSc and PhD projects from the University of Auckland's Leigh Marine Laboratory. He then carried out contract work for New Zealand fisheries and took up a postdoctoral fellowship at the University of Sydney, where in 1988 he obtained a lecturing position in the School of Biological Sciences. Other teaching activities include presenting adult education lectures on fish and marine ecology. His research has focused on reef-associated juvenile and adult fish as well as the early life history stages of fish. He has a major research interest in the importance of oceanographic features for influencing the distribution, movements, and survival of small fish and plankton. He has written many publications on this subject.

JOHN E. McCOSKER

John E. McCosker has been Director of the Steinhart Aquarium, a division of the California Academy of Sciences in San Francisco, since 1973. Although trained as an evolutionary biologist, his research activities have subsequently broadened to include such diverse topics as the symbiotic behavior of bioluminescent fishes, the behavior of venomous sea snakes, the predatory behavior of the great white shark, the biology of penguins, the biology of the coelacanth, and dispersed and renewable energy sources as alternatives to national vulnerability and war. His shark research has been summarized in BBC, NOVA, and *National Geographic* television specials.

COLIN MARTIN

Colin Martin is Reader in the Department of Scottish History at the University of St Andrews, Scotland. He worked as a flying instructor and army officer before becoming a freelance journalist specializing in archeological subjects. Between 1968 and 1983 he was Archeological Director on the excavation of three Spanish Armada wrecks in Irish and Scottish waters. He established the Scottish Institute of Maritime Studies at St Andrews in 1973. He is a frequent broadcaster on radio and television, president of the Nautical Archeology Society, and author of two books on the Spanish Armada and numerous papers and chapters on maritime topics.

JOHN R. PAXTON

John Paxton is a senior research scientist at the Australian Museum, Sydney. He was born in Hollywood and received his degrees from the University of Southern California, working on deep-sea fishes at the Allan Hancock Foundation. He joined the Australian Museum in 1968 as curator of fishes. He has conducted fieldwork in many south Pacific countries and has taken part in numerous deep-sea expeditions. He and coauthors have recently written the first fish volume of the *Zoological Catalogue of Australia* and are working on the second volume. His current work involves attempts to unravel the mysteries of the biology and evolutionary relationships of deep-sea whalefishes.

VICTOR PRESCOTT

Victor Prescott is a political geographer whose first appointment was in Nigeria, at University College Ibadan. While lecturing there for five years he completed his doctoral thesis on Nigeria's international and regional boundaries. Since 1961 he has been on the staff of the Department of Geography at the University of Melbourne. He was appointed to a personal chair in that department in 1986. He is author of *Maritime Political Boundaries of the World* (1986) and *Political Frontiers and Boundaries* (1987).

PAUL SCULLY-POWER

Paul Scully-Power is a research associate at Scripps Institution of Oceanography, University of California, San Diego. He holds the Distinguished Chair of Environmental Acoustics at the Naval Underwater Systems Center, New London, Connecticut, where he is additionally Chief Scientist of the Warfare Systems Architecture and Engineering Office in the Surface ASW Directorate. He is also a flight crew instructor in the Astronaut Office at the Johnson Space Center, Houston, Texas. He has an international reputation in physical oceanography, underwater acoustics, naval systems, and space operations. He was the first Navy civilian and the first oceanographer in space as a crew member of the *Challenger* mission in 1984. He has received the Navy's highest civilian honor, the Distinguished Civilian Service Medal, for his achievements in naval research, and holds over forty Navy special achievement awards.

J.R. SIMONS

J.R. Simons is a former Associate Professor of Biology and Dean of the Faculty of Science at the University of Sydney. After gaining his PhD, he began his academic career as a lecturer in zoology at the University of Sydney. Following his retirement in 1983, he has pursued his interest in the history of Australian biology.

ROBERT E. STEVENSON

Robert E. Stevenson is Secretary-General of the International Association for the Physical Sciences of the Ocean and a consultant to NASA for space oceanography. His distinguished academic career has included teaching and research positions in several universities and consultancies to institutions such as the National Academy of Sciences and NATO. From 1985 to 1988 he was scientific liaison officer and Deputy Director, Space Oceanography, in the Office of Naval Research, Scripps Institution of Oceanography. He is recognized as a leading world authority on space oceanography and continues to work with NASA to develop space oceanography programs.

A.J. UNDERWOOD

A.J. Underwood is Director of the Institute of Marine Ecology and Reader in Experimental Ecology at the University of Sydney. He is experienced in marine ecological investigations in several fields of study. His main interests are the experimental analysis of interactions in complex assemblages of species. He is an authority on the design of sampling and experimental programs and has published numerous scientific papers.

DIANA WALKER

Diana Walker is a senior lecturer in marine botany at the University of Western Australia with particular interests in seagrasses and macroalgae. She was born in the United Kingdom and studied marine biology at the University of Liverpool's Marine Biological Station on the Isle of Man. She carried out her PhD research on coral reef algae in the Red Sea (Jordan) while based at the University of York in England. From there she was appointed to a postdoctoral fellowship and then a lectureship at the University of Western Australia, where she has carried out research on seagrasses and macroalgae from the northwest of Western Australia to the south coast, but concentrating on Shark Bay. Her main area of interest is on the factors influencing the distribution of marine macrophytes, covering ecophysiology to biogeography.

ACKNOWLEDGMENTS

The publishers would like to thank the following people for their assistance in the production of this book: Dr K. Radway Allen, Alistair Barnard (p. 59 illustration), Anne Bowman (p. 42 illustration), Christine Deacon, Frank Knight (p. 65 illustration), Hugh Kirkman, Macquarie Library Pty Ltd, Annette Riddell, Colin Sale, and Trevor Ward. Many of the illustrations prepared for this publication were based on original references provided by the contributors. Other sources of illustrations are listed below.
Page 16 *Stages in continental drift* and *The major tectonic plates* are from *The Macquarie Illustrated World Atlas* 1984, Macquarie Library, Sydney, pages 46–7. They are reproduced by permission of the publishers. **Pages 22–3** *The oceans basins* are reproduced by permission of Verlag Das Beste GmbH, Stuttgart. **Page 28** *The Bering Canyon* is adapted from a block diagram by T.R. Alpha, US Geological Survey. **Page 32** *The ocean currents* is adapted from *The Macquarie Illustrated World Atlas*, page 74. **Pages 38, 40, and 41** Illustrations adapted from A.C. Brown and A. McLachlan, 1990, *Ecology of Sandy Shores*, Elsevier, Amsterdam. **Page 58** *The skipjack tuna* is adapted from *The Times Atlas of the Oceans*, Van Nostrand Reinhold, New York, page 101. **Pages 68–9** *How seabirds catch their prey* is adapted from N.P Ashmole and M.J. Ashmole, 1967, *Comparative Feeding Ecology of Seabirds on a Tropical Oceanic Island*, Yale Peabody Museum Natural History Bulletin 24. **Page 70** *The Arctic tern* is adapted from *The Atlas of the Living World*, 1989, Weidenfeld & Nicolson, London, page 115. **Page 77** *Faunal regions of the Atlantic Ocean* is adapted from R.H. Backus et al, 1977, "Atlantic mesopelagic zoogeography", *Fishes of the Western North Atlantic:Order Iniomi (Myctophiformes)*, Yale University, NY, pp. 266-87. **Page 103** *Divisions of the marine environment* is adapted from H.V. Thurman, 1981, *Introductory Oceanography*, 3rd edn, Charles E. Merrill, Toronto, page 307. **Page 128** *How deep can they go?* is adapted from *National Geographic*, Dec. 1981, page 799. **Page 135** *The global catch* is based on statistics from FAO *Yearbook* 1987. **Page 135** *Using the catch* is used is adapted from N. Myers ed., 1985, *The Gaia Atlas of Planet Management*, Pan Books, London, page 82. **Page 140** *Production stages in typical aquaculture operations* is adapted from C.D. Garland, 1988, *Australian Mariculture: the Role of Hatcheries in Animal Production*, University of Tasmania, Hobart. **Pages 142–3** *Ocean technology* is adapted from *The Gaia Atlas of Planet Management*, pages 78–9. **Page 150** *Limits of discharge the MARPOL convention* is adapted from *The Gaia Atlas of Planet Management*, page 92. **Page 154** *Endangered marine animals* is based on information in *1988–1990 Red List of Threatened Animals*, World Conservation Monitoring Centre, IUCN, Cambridge, 1990.

INDEX